D1562956

THE NATURE OF THINGS

The Nature of Things
Lucretius c. 99 BC – c. 55 BC

Mailing address:
Royal Classics
PO BOX 4608
Main Station Terminal
349 West Georgia Street
Vancouver, BC
Canada, V6B 4A1

Cover design by: A.R. Roumanis

Text set in Minion Pro. Chapter headings set in News Gothic Standard.

ISBN: 978-1-77476-186-1

FIRST EDITION / FIRST PRINTING

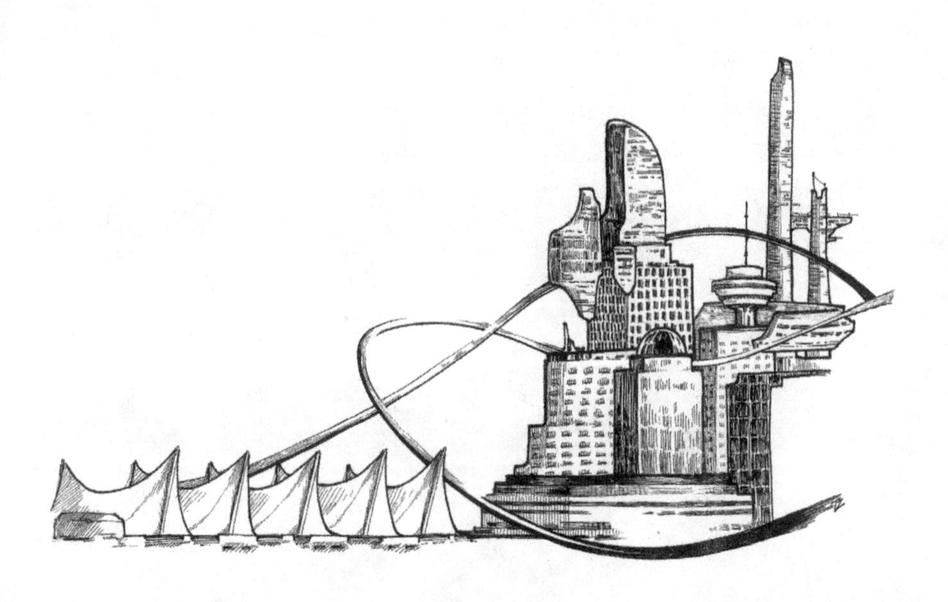

THE NATURE OF THINGS

TITUS LUCRETIUS CARUS

Translated by

William Ellery Leonard

VANCOUVER:
ROYAL CLASSICS
2021

CONTENTS

Book I: Proem 7

- Substance is Eternal 12
- The Void 18
- Nothing Exists per se Except Atoms and the Void 21
- Character of the Atoms 24
- Confutation of Other Philosophers 29
- The Infinity of the Universe 38

Book II: Proem 45

- Atomic Motions 48
- Atomic Forms and their Combinations 57
- Infinite Worlds 79

Book III: Proem 87

- Nature and Composition of the Mind 91
- The Soul is Mortal 101
- Folly of the Fear of Death 115

Book IV: Proem 125

- Existence and Character of the Images 127
- The Senses and Mental Pictures 134
- Some Vital Functions 155
- The Passion of Love 163

Book V: Proem 173

- The World is not Eternal 182
- Origins of Vegetable and Animal Life 203
- Origins and Savage Period of Mankind 209
- Beginnings of Civilization 213

Book VI: Proem 229

- Great Meteorological Phenomena, etc. 233
- The Plague Athens 270

Book One

Proem

Mother of Rome, delight of Gods and men,
Dear Venus that beneath the gliding stars
Makest to teem the many-voyaged main
And fruitful lands–for all of living things
Through thee alone are evermore conceived, 5
Through thee are risen to visit the great sun–
Before thee, Goddess, and thy coming on,
Flee stormy wind and massy cloud away,
For thee the daedal Earth bears scented flowers,
For thee waters of the unvexed deep 10
Smile, and the hollows of the serene sky
Glow with diffused radiance for thee!
For soon as comes the springtime face of day,
And procreant gales blow from the West unbarred,
First fowls of air, smit to the heart by thee, 15
Foretoken thy approach, O thou Divine,
And leap the wild herds round the happy fields
Or swim the bounding torrents. Thus amain,
Seized with the spell, all creatures follow thee
Whithersoever thou walkest forth to lead, 20
And thence through seas and mountains and swift streams,
Through leafy homes of birds and greening plains,
Kindling the lure of love in every breast,
Thou bringest the eternal generations forth,
Kind after kind. And since 'tis thou alone 25
Guidest the Cosmos, and without thee naught

Is risen to reach the shining shores of light,
Nor aught of joyful or of lovely born,
Thee do I crave co-partner in that verse
Which I presume on Nature to compose 30
For Memmius mine, whom thou hast willed to be
Peerless in every grace at every hour–
Wherefore indeed, Divine one, give my words
Immortal charm. Lull to a timely rest
O'er sea and land the savage works of war, 35
For thou alone hast power with public peace
To aid mortality; since he who rules
The savage works of battle, puissant Mars,
How often to thy bosom flings his strength
O'ermastered by the eternal wound of love– 40
And there, with eyes and full throat backward thrown,
Gazing, my Goddess, open-mouthed at thee,
Pastures on love his greedy sight, his breath
Hanging upon thy lips. Him thus reclined
Fill with thy holy body, round, above! 45
Pour from those lips soft syllables to win
Peace for the Romans, glorious Lady, peace!
For in a season troublous to the state
Neither may I attend this task of mine
With thought untroubled, nor mid such events 50
The illustrious scion of the Memmian house
Neglect the civic cause.

Whilst human kind
Throughout the lands lay miserably crushed 55
Before all eyes beneath Religion–who
Would show her head along the region skies,
Glowering on mortals with her hideous face–
A Greek it was who first opposing dared
Raise mortal eyes that terror to withstand, 60
Whom nor the fame of Gods nor lightning's stroke
Nor threatening thunder of the ominous sky
Abashed; but rather chafed to angry zest
His dauntless heart to be the first to rend
The crossbars at the gates of Nature old. 65
And thus his will and hardy wisdom won;
And forward thus he fared afar, beyond

The flaming ramparts of the world, until
He wandered the unmeasurable All.
Whence he to us, a conqueror, reports 70
What things can rise to being, what cannot,
And by what law to each its scope prescribed,
Its boundary stone that clings so deep in Time.
Wherefore Religion now is under foot,
And us his victory now exalts to heaven. 75

I know how hard it is in Latian verse
To tell the dark discoveries of the Greeks,
Chiefly because our pauper-speech must find
Strange terms to fit the strangeness of the thing; 80
Yet worth of thine and the expected joy
Of thy sweet friendship do persuade me on
To bear all toil and wake the clear nights through,
Seeking with what of words and what of song
I may at last most gloriously uncloud 85
For thee the light beyond, wherewith to view
The core of being at the centre hid.
And for the rest, summon to judgments true,
Unbusied ears and singleness of mind
Withdrawn from cares; lest these my gifts, arranged 90
For thee with eager service, thou disdain
Before thou comprehendest: since for thee
I prove the supreme law of Gods and sky,
And the primordial germs of things unfold,
Whence Nature all creates, and multiplies 95
And fosters all, and whither she resolves
Each in the end when each is overthrown.
This ultimate stock we have devised to name
Procreant atoms, matter, seeds of things,
Or primal bodies, as primal to the world. 100

I fear perhaps thou deemest that we fare
An impious road to realms of thought profane;
But 'tis that same religion oftener far
Hath bred the foul impieties of men: 105
As once at Aulis, the elected chiefs,
Foremost of heroes, Danaan counsellors,
Defiled Diana's altar, virgin queen,

With Agamemnon's daughter, foully slain.
She felt the chaplet round her maiden locks 110
And fillets, fluttering down on either cheek,
And at the altar marked her grieving sire,
The priests beside him who concealed the knife,
And all the folk in tears at sight of her.
With a dumb terror and a sinking knee 115
She dropped; nor might avail her now that first
'Twas she who gave the king a father's name.
They raised her up, they bore the trembling girl
On to the altar–hither led not now
With solemn rites and hymeneal choir, 120
But sinless woman, sinfully foredone,
A parent felled her on her bridal day,
Making his child a sacrificial beast
To give the ships auspicious winds for Troy:
Such are the crimes to which Religion leads. 125

And there shall come the time when even thou,
Forced by the soothsayer's terror-tales, shalt seek
To break from us. Ah, many a dream even now
Can they concoct to rout thy plans of life, 130
And trouble all thy fortunes with base fears.
I own with reason: for, if men but knew
Some fixed end to ills, they would be strong
By some device unconquered to withstand
Religions and the menacings of seers. 135
But now nor skill nor instrument is theirs,
Since men must dread eternal pains in death.
For what the soul may be they do not know,
Whether 'tis born, or enter in at birth,
And whether, snatched by death, it die with us, 140
Or visit the shadows and the vasty caves
Of Orcus, or by some divine decree
Enter the brute herds, as our Ennius sang,
Who first from lovely Helicon brought down
A laurel wreath of bright perennial leaves, 145
Renowned forever among the Italian clans.
Yet Ennius too in everlasting verse
Proclaims those vaults of Acheron to be,
Though thence, he said, nor souls nor bodies fare,

But only phantom figures, strangely wan, 150
And tells how once from out those regions rose
Old Homer's ghost to him and shed salt tears
And with his words unfolded Nature's source.
Then be it ours with steady mind to clasp
The purport of the skies–the law behind 155
The wandering courses of the sun and moon;
To scan the powers that speed all life below;
But most to see with reasonable eyes
Of what the mind, of what the soul is made,
And what it is so terrible that breaks 160
On us asleep, or waking in disease,
Until we seem to mark and hear at hand
Dead men whose bones earth bosomed long ago.

Substance is Eternal

This terror, then, this darkness of the mind,
Not sunrise with its flaring spokes of light,
Nor glittering arrows of morning can disperse,
But only Nature's aspect and her law,
Which, teaching us, hath this exordium: 5
Nothing from nothing ever yet was born.
Fear holds dominion over mortality
Only because, seeing in land and sky
So much the cause whereof no wise they know,
Men think Divinities are working there. 10
Meantime, when once we know from nothing still
Nothing can be create, we shall divine
More clearly what we seek: those elements
From which alone all things created are,
And how accomplished by no tool of Gods. 15
Suppose all sprang from all things: any kind
Might take its origin from any thing,
No fixed seed required. Men from the sea
Might rise, and from the land the scaly breed,
And, fowl full fledged come bursting from the sky; 20
The horned cattle, the herds and all the wild
Would haunt with varying offspring tilth and waste;
Nor would the same fruits keep their olden trees,
But each might grow from any stock or limb
By chance and change. Indeed, and were there not 25
For each its procreant atoms, could things have
Each its unalterable mother old?
But, since produced from fixed seeds are all,
Each birth goes forth upon the shores of light

From its own stuff, from its own primal bodies. 30
And all from all cannot become, because
In each resides a secret power its own.
Again, why see we lavished o'er the lands
At spring the rose, at summer heat the corn,
The vines that mellow when the autumn lures, 35
If not because the fixed seeds of things
At their own season must together stream,
And new creations only be revealed
When the due times arrive and pregnant earth
Safely may give unto the shores of light 40
Her tender progenies? But if from naught
Were their becoming, they would spring abroad
Suddenly, unforeseen, in alien months,
With no primordial germs, to be preserved
From procreant unions at an adverse hour. 45
Nor on the mingling of the living seeds
Would space be needed for the growth of things
Were life an increment of nothing: then
The tiny babe forthwith would walk a man,
And from the turf would leap a branching tree– 50
Wonders unheard of; for, by Nature, each
Slowly increases from its lawful seed,
And through that increase shall conserve its kind.
Whence take the proof that things enlarge and feed
From out their proper matter. Thus it comes 55
That earth, without her seasons of fixed rains,
Could bear no produce such as makes us glad,
And whatsoever lives, if shut from food,
Prolongs its kind and guards its life no more.
Thus easier 'tis to hold that many things 60
Have primal bodies in common (as we see
The single letters common to many words)
Than aught exists without its origins.
Moreover, why should Nature not prepare
Men of a bulk to ford the seas afoot, 65
Or rend the mighty mountains with their hands,
Or conquer Time with length of days, if not
Because for all begotten things abides
The changeless stuff, and what from that may spring
Is fixed forevermore? Lastly we see 70

How far the tilled surpass the fields untilled
And to the labour of our hands return
Their more abounding crops; there are indeed
Within the earth primordial germs of things,
Which, as the ploughshare turns the fruitful clods 75
And kneads the mould, we quicken into birth.
Else would ye mark, without all toil of ours,
Spontaneous generations, fairer forms.
Confess then, naught from nothing can become,
Since all must have their seeds, wherefrom to grow, 80
Wherefrom to reach the gentle fields of air.
Hence too it comes that Nature all dissolves
Into their primal bodies again, and naught
Perishes ever to annihilation.
For, were aught mortal in its every part, 85
Before our eyes it might be snatched away
Unto destruction; since no force were needed
To sunder its members and undo its bands.
Whereas, of truth, because all things exist,
With seed imperishable, Nature allows 90
Destruction nor collapse of aught, until
Some outward force may shatter by a blow,
Or inward craft, entering its hollow cells,
Dissolve it down. And more than this, if Time,
That wastes with eld the works along the world, 95
Destroy entire, consuming matter all,
Whence then may Venus back to light of life
Restore the generations kind by kind?
Or how, when thus restored, may daedal Earth
Foster and plenish with her ancient food, 100
Which, kind by kind, she offers unto each?
Whence may the water-springs, beneath the sea,
Or inland rivers, far and wide away,
Keep the unfathomable ocean full?
And out of what does Ether feed the stars? 105
For lapsed years and infinite age must else
Have eat all shapes of mortal stock away:
But be it the Long Ago contained those germs,
By which this sum of things recruited lives,
Those same infallibly can never die, 110
Nor nothing to nothing evermore return.

And, too, the selfsame power might end alike
All things, were they not still together held
By matter eternal, shackled through its parts,
Now more, now less. A touch might be enough 115
To cause destruction. For the slightest force
Would loose the weft of things wherein no part
Were of imperishable stock. But now
Because the fastenings of primordial parts
Are put together diversely and stuff 120
Is everlasting, things abide the same
Unhurt and sure, until some power comes on
Strong to destroy the warp and woof of each:
Nothing returns to naught; but all return
At their collapse to primal forms of stuff. 125
Lo, the rains perish which Ether-father throws
Down to the bosom of Earth-mother; but then
Upsprings the shining grain, and boughs are green
Amid the trees, and trees themselves wax big
And lade themselves with fruits; and hence in turn 130
The race of man and all the wild are fed;
Hence joyful cities thrive with boys and girls;
And leafy woodlands echo with new birds;
Hence cattle, fat and drowsy, lay their bulk
Along the joyous pastures whilst the drops 135
Of white ooze trickle from distended bags;
Hence the young scamper on their weakling joints
Along the tender herbs, fresh hearts afrisk
With warm new milk. Thus naught of what so seems
Perishes utterly, since Nature ever 140
Upbuilds one thing from other, suffering naught
To come to birth but through some other's death.

And now, since I have taught that things cannot
Be born from nothing, nor the same, when born, 145
To nothing be recalled, doubt not my words,
Because our eyes no primal germs perceive;
For mark those bodies which, though known to be
In this our world, are yet invisible:
The winds infuriate lash our face and frame, 150
Unseen, and swamp huge ships and rend the clouds,
Or, eddying wildly down, bestrew the plains

With mighty trees, or scour the mountain tops
With forest-crackling blasts. Thus on they rave
With uproar shrill and ominous moan. The winds, 155
'Tis clear, are sightless bodies sweeping through
The sea, the lands, the clouds along the sky,
Vexing and whirling and seizing all amain;
And forth they flow and pile destruction round,
Even as the water's soft and supple bulk 160
Becoming a river of abounding floods,
Which a wide downpour from the lofty hills
Swells with big showers, dashes headlong down
Fragments of woodland and whole branching trees;
Nor can the solid bridges bide the shock 165
As on the waters whelm: the turbulent stream,
Strong with a hundred rains, beats round the piers,
Crashes with havoc, and rolls beneath its waves
Down-toppled masonry and ponderous stone,
Hurling away whatever would oppose. 170
Even so must move the blasts of all the winds,
Which, when they spread, like to a mighty flood,
Hither or thither, drive things on before
And hurl to ground with still renewed assault,
Or sometimes in their circling vortex seize 175
And bear in cones of whirlwind down the world:
The winds are sightless bodies and naught else–
Since both in works and ways they rival well
The mighty rivers, the visible in form.
Then too we know the varied smells of things 180
Yet never to our nostrils see them come;
With eyes we view not burning heats, nor cold,
Nor are we wont men's voices to behold.
Yet these must be corporeal at the base,
Since thus they smite the senses: naught there is 185
Save body, having property of touch.
And raiment, hung by surf-beat shore, grows moist,
The same, spread out before the sun, will dry;
Yet no one saw how sank the moisture in,
Nor how by heat off-driven. Thus we know, 190
That moisture is dispersed about in bits
Too small for eyes to see. Another case:
A ring upon the finger thins away

Along the under side, with years and suns;
The drippings from the eaves will scoop the stone; 195
The hooked ploughshare, though of iron, wastes
Amid the fields insidiously. We view
The rock-paved highways worn by many feet;
And at the gates the brazen statues show
Their right hands leaner from the frequent touch 200
Of wayfarers innumerable who greet.
We see how wearing-down hath minished these,
But just what motes depart at any time,
The envious nature of vision bars our sight.
Lastly whatever days and nature add 205
Little by little, constraining things to grow
In due proportion, no gaze however keen
Of these our eyes hath watched and known. No more
Can we observe what's lost at any time,
When things wax old with eld and foul decay, 210
Or when salt seas eat under beetling crags.
Thus Nature ever by unseen bodies works.

The Void

But yet creation's neither crammed nor blocked
About by body: there's in things a void–
Which to have known will serve thee many a turn,
Nor will not leave thee wandering in doubt,
Forever searching in the sum of all, 5
And losing faith in these pronouncements mine.
There's place intangible, a void and room.
For were it not, things could in nowise move;
Since body's property to block and check
Would work on all and at an times the same. 10
Thus naught could evermore push forth and go,
Since naught elsewhere would yield a starting place.
But now through oceans, lands, and heights of heaven,
By divers causes and in divers modes,
Before our eyes we mark how much may move, 15
Which, finding not a void, would fail deprived
Of stir and motion; nay, would then have been
Nowise begot at all, since matter, then,
Had staid at rest, its parts together crammed.
Then too, however solid objects seem, 20
They yet are formed of matter mixed with void:
In rocks and caves the watery moisture seeps,
And beady drops stand out like plenteous tears;
And food finds way through every frame that lives;
The trees increase and yield the season's fruit 25
Because their food throughout the whole is poured,
Even from the deepest roots, through trunks and boughs;
And voices pass the solid walls and fly
Reverberant through shut doorways of a house;

And stiffening frost seeps inward to our bones. 30
Which but for voids for bodies to go through
'Tis clear could happen in nowise at all.
Again, why see we among objects some
Of heavier weight, but of no bulkier size?
Indeed, if in a ball of wool there be 35
As much of body as in lump of lead,
The two should weigh alike, since body tends
To load things downward, while the void abides,
By contrary nature, the imponderable.
Therefore, an object just as large but lighter 40
Declares infallibly its more of void;
Even as the heavier more of matter shows,
And how much less of vacant room inside.
That which we're seeking with sagacious quest
Exists, infallibly, commixed with things– 45
The void, the invisible inane.

Right here
I am compelled a question to expound,
Forestalling something certain folk suppose, 50
Lest it avail to lead thee off from truth:
Waters (they say) before the shining breed
Of the swift scaly creatures somehow give,
And straightway open sudden liquid paths,
Because the fishes leave behind them room 55
To which at once the yielding billows stream.
Thus things among themselves can yet be moved,
And change their place, however full the Sum–
Received opinion, wholly false forsooth.
For where can scaly creatures forward dart, 60
Save where the waters give them room? Again,
Where can the billows yield a way, so long
As ever the fish are powerless to go?
Thus either all bodies of motion are deprived,
Or things contain admixture of a void 65
Where each thing gets its start in moving on.

Lastly, where after impact two broad bodies
Suddenly spring apart, the air must crowd
The whole new void between those bodies formed; 70

But air, however it stream with hastening gusts,
Can yet not fill the gap at once–for first
It makes for one place, ere diffused through all.
And then, if haply any think this comes,
When bodies spring apart, because the air 75
Somehow condenses, wander they from truth:
For then a void is formed, where none before;
And, too, a void is filled which was before.
Nor can air be condensed in such a wise;
Nor, granting it could, without a void, I hold, 80
It still could not contract upon itself
And draw its parts together into one.
Wherefore, despite demur and counter-speech,
Confess thou must there is a void in things.

85

And still I might by many an argument
Here scrape together credence for my words.
But for the keen eye these mere footprints serve,
Whereby thou mayest know the rest thyself.
As dogs full oft with noses on the ground, 90
Find out the silent lairs, though hid in brush,
Of beasts, the mountain-rangers, when but once
They scent the certain footsteps of the way,
Thus thou thyself in themes like these alone
Can hunt from thought to thought, and keenly wind 95
Along even onward to the secret places
And drag out truth. But, if thou loiter loth
Or veer, however little, from the point,
This I can promise, Memmius, for a fact:
Such copious drafts my singing tongue shall pour 100
From the large well-springs of my plenished breast
That much I dread slow age will steal and coil
Along our members, and unloose the gates
Of life within us, ere for thee my verse
Hath put within thine ears the stores of proofs 105
At hand for one soever question broached.

Nothing Exists per se Except Atoms and the Void

But, now again to weave the tale begun,
All nature, then, as self-sustained, consists
Of twain of things: of bodies and of void
In which they're set, and where they're moved around.
For common instinct of our race declares 5
That body of itself exists: unless
This primal faith, deep-founded, fail us not,
Naught will there be whereunto to appeal
On things occult when seeking aught to prove
By reasonings of mind. Again, without 10
That place and room, which we do call the inane,
Nowhere could bodies then be set, nor go
Hither or thither at all–as shown before.
Besides, there's naught of which thou canst declare
It lives disjoined from body, shut from void– 15
A kind of third in nature. For whatever
Exists must be a somewhat; and the same,
If tangible, however fight and slight,
Will yet increase the count of body's sum,
With its own augmentation big or small; 20
But, if intangible and powerless ever
To keep a thing from passing through itself
On any side, 'twill be naught else but that
Which we do call the empty, the inane.
Again, whate'er exists, as of itself, 25
Must either act or suffer action on it,
Or else be that wherein things move and be:

Naught, saving body, acts, is acted on;
Naught but the inane can furnish room. And thus,
Beside the inane and bodies, is no third 30
Nature amid the number of all things–
Remainder none to fall at any time
Under our senses, nor be seized and seen
By any man through reasonings of mind.
Name o'er creation with what names thou wilt, 35
Thou'lt find but properties of those first twain,
Or see but accidents those twain produce.

A property is that which not at all
Can be disjoined and severed from a thing 40
Without a fatal dissolution: such,
Weight to the rocks, heat to the fire, and flow
To the wide waters, touch to corporal things,
Intangibility to the viewless void.
But state of slavery, pauperhood, and wealth, 45
Freedom, and war, and concord, and all else
Which come and go whilst nature stands the same,
We're wont, and rightly, to call accidents.
Even time exists not of itself; but sense
Reads out of things what happened long ago, 50
What presses now, and what shall follow after:
No man, we must admit, feels time itself,
Disjoined from motion and repose of things.
Thus, when they say there "is" the ravishment
Of Princess Helen, "is" the siege and sack 55
Of Trojan Town, look out, they force us not
To admit these acts existent by themselves,
Merely because those races of mankind
(Of whom these acts were accidents) long since
Irrevocable age has borne away: 60
For all past actions may be said to be
But accidents, in one way, of mankind,–
In other, of some region of the world.
Add, too, had been no matter, and no room
Wherein all things go on, the fire of love 65
Upblown by that fair form, the glowing coal
Under the Phrygian Alexander's breast,
Had ne'er enkindled that renowned strife

Of savage war, nor had the wooden horse
Involved in flames old Pergama, by a birth 70
At midnight of a brood of the Hellenes.
And thus thou canst remark that every act
At bottom exists not of itself, nor is
As body is, nor has like name with void;
But rather of sort more fitly to be called 75
An accident of body, and of place
Wherein all things go on.

Character of the Atoms

Bodies, again,
Are partly primal germs of things, and partly
Unions deriving from the primal germs.
And those which are the primal germs of things
No power can quench; for in the end they conquer 5
By their own solidness; though hard it be
To think that aught in things has solid frame;
For lightnings pass, no less than voice and shout,
Through hedging walls of houses, and the iron
White-dazzles in the fire, and rocks will burn 10
With exhalations fierce and burst asunder.
Totters the rigid gold dissolved in heat;
The ice of bronze melts conquered in the flame;
Warmth and the piercing cold through silver seep,
Since, with the cups held rightly in the hand, 15
We oft feel both, as from above is poured
The dew of waters between their shining sides:
So true it is no solid form is found.
But yet because true reason and nature of things
Constrain us, come, whilst in few verses now 20
I disentangle how there still exist
Bodies of solid, everlasting frame–
The seeds of things, the primal germs we teach,
Whence all creation around us came to be.
First since we know a twofold nature exists, 25
Of things, both twain and utterly unlike–
Body, and place in which an things go on–
Then each must be both for and through itself,
And all unmixed: where'er be empty space,

There body's not; and so where body bides, 30
There not at all exists the void inane.
Thus primal bodies are solid, without a void.
But since there's void in all begotten things,
All solid matter must be round the same;
Nor, by true reason canst thou prove aught hides 35
And holds a void within its body, unless
Thou grant what holds it be a solid. Know,
That which can hold a void of things within
Can be naught else than matter in union knit.
Thus matter, consisting of a solid frame, 40
Hath power to be eternal, though all else,
Though all creation, be dissolved away.
Again, were naught of empty and inane,
The world were then a solid; as, without
Some certain bodies to fill the places held, 45
The world that is were but a vacant void.
And so, infallibly, alternate-wise
Body and void are still distinguished,
Since nature knows no wholly full nor void.
There are, then, certain bodies, possessed of power 50
To vary forever the empty and the full;
And these can nor be sundered from without
By beats and blows, nor from within be torn
By penetration, nor be overthrown
By any assault soever through the world– 55
For without void, naught can be crushed, it seems,
Nor broken, nor severed by a cut in twain,
Nor can it take the damp, or seeping cold
Or piercing fire, those old destroyers three;
But the more void within a thing, the more 60
Entirely it totters at their sure assault.
Thus if first bodies be, as I have taught,
Solid, without a void, they must be then
Eternal; and, if matter ne'er had been
Eternal, long ere now had all things gone 65
Back into nothing utterly, and all
We see around from nothing had been born–
But since I taught above that naught can be
From naught created, nor the once begotten
To naught be summoned back, these primal germs 70

Must have an immortality of frame.
And into these must each thing be resolved,
When comes its supreme hour, that thus there be
At hand the stuff for plenishing the world.

75

So primal germs have solid singleness
Nor otherwise could they have been conserved
Through aeons and infinity of time
For the replenishment of wasted worlds.
Once more, if nature had given a scope for things 80
To be forever broken more and more,
By now the bodies of matter would have been
So far reduced by breakings in old days
That from them nothing could, at season fixed,
Be born, and arrive its prime and top of life. 85
For, lo, each thing is quicker marred than made;
And so whate'er the long infinitude
Of days and all fore-passed time would now
By this have broken and ruined and dissolved,
That same could ne'er in all remaining time 90
Be builded up for plenishing the world.
But mark: infallibly a fixed bound
Remaineth stablished 'gainst their breaking down;
Since we behold each thing soever renewed,
And unto all, their seasons, after their kind, 95
Wherein they arrive the flower of their age.

Again, if bounds have not been set against
The breaking down of this corporeal world,
Yet must all bodies of whatever things 100
Have still endured from everlasting time
Unto this present, as not yet assailed
By shocks of peril. But because the same
Are, to thy thinking, of a nature frail,
It ill accords that thus they could remain 105
(As thus they do) through everlasting time,
Vexed through the ages (as indeed they are)
By the innumerable blows of chance.

So in our programme of creation, mark 110
How 'tis that, though the bodies of all stuff

Are solid to the core, we yet explain
The ways whereby some things are fashioned soft–
Air, water, earth, and fiery exhalations–
And by what force they function and go on: 115
The fact is founded in the void of things.
But if the primal germs themselves be soft,
Reason cannot be brought to bear to show
The ways whereby may be created these
Great crags of basalt and the during iron; 120
For their whole nature will profoundly lack
The first foundations of a solid frame.
But powerful in old simplicity,
Abide the solid, the primeval germs;
And by their combinations more condensed, 125
All objects can be tightly knit and bound
And made to show unconquerable strength.
Again, since all things kind by kind obtain
Fixed bounds of growing and conserving life;
Since Nature hath inviolably decreed 130
What each can do, what each can never do;
Since naught is changed, but all things so abide
That ever the variegated birds reveal
The spots or stripes peculiar to their kind,
Spring after spring: thus surely all that is 135
Must be composed of matter immutable.
For if the primal germs in any wise
Were open to conquest and to change, 'twould be
Uncertain also what could come to birth
And what could not, and by what law to each 140
Its scope prescribed, its boundary stone that clings
So deep in Time. Nor could the generations
Kind after kind so often reproduce
The nature, habits, motions, ways of life,
Of their progenitors. 145

And then again,
Since there is ever an extreme bounding point
Of that first body which our senses now
Cannot perceive: That bounding point indeed 150
Exists without all parts, a minimum
Of nature, nor was e'er a thing apart,

As of itself,–nor shall hereafter be,
Since 'tis itself still parcel of another,
A first and single part, whence other parts 155
And others similar in order lie
In a packed phalanx, filling to the full
The nature of first body: being thus
Not self-existent, they must cleave to that
From which in nowise they can sundered be. 160
So primal germs have solid singleness,
Which tightly packed and closely joined cohere
By virtue of their minim particles–
No compound by mere union of the same;
But strong in their eternal singleness, 165
Nature, reserving them as seeds for things,
Permitteth naught of rupture or decrease.

Moreover, were there not a minimum,
The smallest bodies would have infinites, 170
Since then a half-of-half could still be halved,
With limitless division less and less.
Then what the difference 'twixt the sum and least?
None: for however infinite the sum,
Yet even the smallest would consist the same 175
Of infinite parts. But since true reason here
Protests, denying that the mind can think it,
Convinced thou must confess such things there are
As have no parts, the minimums of nature.
And since these are, likewise confess thou must 180
That primal bodies are solid and eterne.
Again, if Nature, creatress of all things,
Were wont to force all things to be resolved
Unto least parts, then would she not avail
To reproduce from out them anything; 185
Because whate'er is not endowed with parts
Cannot possess those properties required
Of generative stuff–divers connections,
Weights, blows, encounters, motions, whereby things
Forevermore have being and go on. 190

Confutation of Other Philosophers

And on such grounds it is that those who held
The stuff of things is fire, and out of fire
Alone the cosmic sum is formed, are seen
Mightily from true reason to have lapsed.
Of whom, chief leader to do battle, comes 5
That Heraclitus, famous for dark speech
Among the silly, not the serious Greeks
Who search for truth. For dolts are ever prone
That to bewonder and adore which hides
Beneath distorted words, holding that true 10
Which sweetly tickles in their stupid ears,
Or which is rouged in finely finished phrase.
For how, I ask, can things so varied be,
If formed of fire, single and pure? No whit
'Twould help for fire to be condensed or thinned, 15
If all the parts of fire did still preserve
But fire's own nature, seen before in gross.
The heat were keener with the parts compressed,
Milder, again, when severed or dispersed–
And more than this thou canst conceive of naught 20
That from such causes could become; much less
Might earth's variety of things be born
From any fires soever, dense or rare.
This too: if they suppose a void in things,
Then fires can be condensed and still left rare; 25
But since they see such opposites of thought
Rising against them, and are loath to leave
An unmixed void in things, they fear the steep
And lose the road of truth. Nor do they see,

That, if from things we take away the void, 30
All things are then condensed, and out of all
One body made, which has no power to dart
Swiftly from out itself not anything–
As throws the fire its light and warmth around,
Giving thee proof its parts are not compact. 35
But if perhaps they think, in other wise,
Fires through their combinations can be quenched
And change their substance, very well: behold,
If fire shall spare to do so in no part,
Then heat will perish utterly and all, 40
And out of nothing would the world be formed.
For change in anything from out its bounds
Means instant death of that which was before;
And thus a somewhat must persist unharmed
Amid the world, lest all return to naught, 45
And, born from naught, abundance thrive anew.
Now since indeed there are those surest bodies
Which keep their nature evermore the same,
Upon whose going out and coming in
And changed order things their nature change, 50
And all corporeal substances transformed,
'Tis thine to know those primal bodies, then,
Are not of fire. For 'twere of no avail
Should some depart and go away, and some
Be added new, and some be changed in order, 55
If still all kept their nature of old heat:
For whatsoever they created then
Would still in any case be only fire.
The truth, I fancy, this: bodies there are
Whose clashings, motions, order, posture, shapes 60
Produce the fire and which, by order changed,
Do change the nature of the thing produced,
And are thereafter nothing like to fire
Nor whatso else has power to send its bodies
With impact touching on the senses' touch. 65

Again, to say that all things are but fire
And no true thing in number of all things
Exists but fire, as this same fellow says,
Seems crazed folly. For the man himself 70

Against the senses by the senses fights,
And hews at that through which is all belief,
Through which indeed unto himself is known
The thing he calls the fire. For, though he thinks
The senses truly can perceive the fire, 75
He thinks they cannot as regards all else,
Which still are palpably as clear to sense–
To me a thought inept and crazy too.
For whither shall we make appeal? for what
More certain than our senses can there be 80
Whereby to mark asunder error and truth?
Besides, why rather do away with all,
And wish to allow heat only, then deny
The fire and still allow all else to be?–
Alike the madness either way it seems. 85
Thus whosoe'er have held the stuff of things
To be but fire, and out of fire the sum,
And whosoever have constituted air
As first beginning of begotten things,
And all whoever have held that of itself 90
Water alone contrives things, or that earth
Createth all and changes things anew
To divers natures, mightily they seem
A long way to have wandered from the truth.
95

Add, too, whoever make the primal stuff
Twofold, by joining air to fire, and earth
To water; add who deem that things can grow
Out of the four–fire, earth, and breath, and rain;
As first Empedocles of Acragas, 100
Whom that three-cornered isle of all the lands
Bore on her coasts, around which flows and flows
In mighty bend and bay the Ionic seas,
Splashing the brine from off their gray-green waves.
Here, billowing onward through the narrow straits, 105
Swift ocean cuts her boundaries from the shores
Of the Italic mainland. Here the waste
Charybdis; and here Aetna rumbles threats
To gather anew such furies of its flames
As with its force anew to vomit fires, 110
Belched from its throat, and skyward bear anew

Its lightnings' flash. And though for much she seem
The mighty and the wondrous isle to men,
Most rich in all good things, and fortified
With generous strength of heroes, she hath ne'er 115
Possessed within her aught of more renown,
Nor aught more holy, wonderful, and dear
Than this true man. Nay, ever so far and pure
The lofty music of his breast divine
Lifts up its voice and tells of glories found, 120
That scarce he seems of human stock create.

Yet he and those forementioned (known to be
So far beneath him, less than he in all),
Though, as discoverers of much goodly truth, 125
They gave, as 'twere from out of the heart's own shrine,
Responses holier and soundlier based
Than ever the Pythia pronounced for men
From out the triped and the Delphian laurel,
Have still in matter of first-elements 130
Made ruin of themselves, and, great men, great
Indeed and heavy there for them the fall:
First, because, banishing the void from things,
They yet assign them motion, and allow
Things soft and loosely textured to exist, 135
As air, dew, fire, earth, animals, and grains,
Without admixture of void amid their frame.
Next, because, thinking there can be no end
In cutting bodies down to less and less
Nor pause established to their breaking up, 140
They hold there is no minimum in things;
Albeit we see the boundary point of aught
Is that which to our senses seems its least,
Whereby thou mayst conjecture, that, because
The things thou canst not mark have boundary points, 145
They surely have their minimums. Then, too,
Since these philosophers ascribe to things
Soft primal germs, which we behold to be
Of birth and body mortal, thus, throughout,
The sum of things must be returned to naught, 150
And, born from naught, abundance thrive anew–
Thou seest how far each doctrine stands from truth.

And, next, these bodies are among themselves
In many ways poisons and foes to each,
Wherefore their congress will destroy them quite 155
Or drive asunder as we see in storms
Rains, winds, and lightnings all asunder fly.

Thus too, if all things are create of four,
And all again dissolved into the four, 160
How can the four be called the primal germs
Of things, more than all things themselves be thought,
By retroversion, primal germs of them?
For ever alternately are both begot,
With interchange of nature and aspect 165
From immemorial time. But if percase
Thou think'st the frame of fire and earth, the air,
The dew of water can in such wise meet
As not by mingling to resign their nature,
From them for thee no world can be create– 170
No thing of breath, no stock or stalk of tree:
In the wild congress of this varied heap
Each thing its proper nature will display,
And air will palpably be seen mixed up
With earth together, unquenched heat with water. 175
But primal germs in bringing things to birth
Must have a latent, unseen quality,
Lest some outstanding alien element
Confuse and minish in the thing create
Its proper being. 180

But these men begin
From heaven, and from its fires; and first they feign
That fire will turn into the winds of air,
Next, that from air the rain begotten is, 185
And earth created out of rain, and then
That all, reversely, are returned from earth–
The moisture first, then air thereafter heat–
And that these same ne'er cease in interchange,
To go their ways from heaven to earth, from earth 190
Unto the stars of the aethereal world–
Which in no wise at all the germs can do.
Since an immutable somewhat still must be,

Lest all things utterly be sped to naught;
For change in anything from out its bounds 195
Means instant death of that which was before.
Wherefore, since those things, mentioned heretofore,
Suffer a changed state, they must derive
From others ever unconvertible,
Lest an things utterly return to naught. 200
Then why not rather presuppose there be
Bodies with such a nature furnished forth
That, if perchance they have created fire,
Can still (by virtue of a few withdrawn,
Or added few, and motion and order changed) 205
Fashion the winds of air, and thus all things
Forevermore be interchanged with all?

"But facts in proof are manifest," thou sayest,
"That all things grow into the winds of air 210
And forth from earth are nourished, and unless
The season favour at propitious hour
With rains enough to set the trees a-reel
Under the soak of bulking thunderheads,
And sun, for its share, foster and give heat, 215
No grains, nor trees, nor breathing things can grow."
True–and unless hard food and moisture soft
Recruited man, his frame would waste away,
And life dissolve from out his thews and bones;
For out of doubt recruited and fed are we 220
By certain things, as other things by others.
Because in many ways the many germs
Common to many things are mixed in things,
No wonder 'tis that therefore divers things
By divers things are nourished. And, again, 225
Often it matters vastly with what others,
In what positions the primordial germs
Are bound together, and what motions, too,
They give and get among themselves; for these
Same germs do put together sky, sea, lands, 230
Rivers, and sun, grains, trees, and breathing things,
But yet commixed they are in divers modes
With divers things, forever as they move.
Nay, thou beholdest in our verses here

Elements many, common to many worlds, 235
Albeit thou must confess each verse, each word
From one another differs both in sense
And ring of sound–so much the elements
Can bring about by change of order alone.
But those which are the primal germs of things 240
Have power to work more combinations still,
Whence divers things can be produced in turn.

Now let us also take for scrutiny
The homeomeria of Anaxagoras, 245
So called by Greeks, for which our pauper-speech
Yieldeth no name in the Italian tongue,
Although the thing itself is not o'erhard
For explanation. First, then, when he speaks
Of this homeomeria of things, he thinks 250
Bones to be sprung from littlest bones minute,
And from minute and littlest flesh all flesh,
And blood created out of drops of blood,
Conceiving gold compact of grains of gold,
And earth concreted out of bits of earth, 255
Fire made of fires, and water out of waters,
Feigning the like with all the rest of stuff.
Yet he concedes not any void in things,
Nor any limit to cutting bodies down.
Wherefore to me he seems on both accounts 260
To err no less than those we named before.
Add too: these germs he feigns are far too frail–
If they be germs primordial furnished forth
With but same nature as the things themselves,
And travail and perish equally with those, 265
And no rein curbs them from annihilation.
For which will last against the grip and crush
Under the teeth of death? the fire? the moist?
Or else the air? which then? the blood? the bones?
No one, methinks, when every thing will be 270
At bottom as mortal as whate'er we mark
To perish by force before our gazing eyes.
But my appeal is to the proofs above
That things cannot fall back to naught, nor yet
From naught increase. And now again, since food 275

Augments and nourishes the human frame,
'Tis thine to know our veins and blood and bones
And thews are formed of particles unlike
To them in kind; or if they say all foods
Are of mixed substance having in themselves 280
Small bodies of thews, and bones, and also veins
And particles of blood, then every food,
Solid or liquid, must itself be thought
As made and mixed of things unlike in kind–
Of bones, of thews, of ichor and of blood. 285
Again, if all the bodies which upgrow
From earth, are first within the earth, then earth
Must be compound of alien substances.
Which spring and bloom abroad from out the earth.
Transfer the argument, and thou may'st use 290
The selfsame words: if flame and smoke and ash
Still lurk unseen within the wood, the wood
Must be compound of alien substances
Which spring from out the wood.

295

Right here remains
A certain slender means to skulk from truth,
Which Anaxagoras takes unto himself,
Who holds that all things lurk commixed with all
While that one only comes to view, of which 300
The bodies exceed in number all the rest,
And lie more close to hand and at the fore–
A notion banished from true reason far.
For then 'twere meet that kernels of the grains
Should oft, when crunched between the might of stones, 305
Give forth a sign of blood, or of aught else
Which in our human frame is fed; and that
Rock rubbed on rock should yield a gory ooze.
Likewise the herbs ought oft to give forth drops
Of sweet milk, flavoured like the uddered sheep's; 310
Indeed we ought to find, when crumbling up
The earthy clods, there herbs, and grains, and leaves,
All sorts dispersed minutely in the soil;
Lastly we ought to find in cloven wood
Ashes and smoke and bits of fire there hid. 315
But since fact teaches this is not the case,

'Tis thine to know things are not mixed with things
Thuswise; but seeds, common to many things,
Commixed in many ways, must lurk in things.

320

"But often it happens on skiey hills" thou sayest,
"That neighbouring tops of lofty trees are rubbed
One against other, smote by the blustering south,
Till all ablaze with bursting flower of flame."
Good sooth–yet fire is not ingraft in wood, 325
But many are the seeds of heat, and when
Rubbing together they together flow,
They start the conflagrations in the forests.
Whereas if flame, already fashioned, lay
Stored up within the forests, then the fires 330
Could not for any time be kept unseen,
But would be laying all the wildwood waste
And burning all the boscage. Now dost see
(Even as we said a little space above)
How mightily it matters with what others, 335
In what positions these same primal germs
Are bound together? And what motions, too,
They give and get among themselves? how, hence,
The same, if altered 'mongst themselves, can body
Both igneous and ligneous objects forth– 340
Precisely as these words themselves are made
By somewhat altering their elements,
Although we mark with name indeed distinct
The igneous from the ligneous. Once again,
If thou suppose whatever thou beholdest, 345
Among all visible objects, cannot be,
Unless thou feign bodies of matter endowed
With a like nature,–by thy vain device
For thee will perish all the germs of things:
'Twill come to pass they'll laugh aloud, like men, 350
Shaken asunder by a spasm of mirth,
Or moisten with salty tear-drops cheeks and chins.

The Infinity of the Universe

Now learn of what remains! More keenly hear!
And for myself, my mind is not deceived
How dark it is: But the large hope of praise
Hath strook with pointed thyrsus through my heart;
On the same hour hath strook into my breast 5
Sweet love of the Muses, wherewith now instinct,
I wander afield, thriving in sturdy thought,
Through unpathed haunts of the Pierides,
Trodden by step of none before. I joy
To come on undefiled fountains there, 10
To drain them deep; I joy to pluck new flowers,
To seek for this my head a signal crown
From regions where the Muses never yet
Have garlanded the temples of a man:
First, since I teach concerning mighty things, 15
And go right on to loose from round the mind
The tightened coils of dread religion;
Next, since, concerning themes so dark, I frame
Songs so pellucid, touching all throughout
Even with the Muses' charm–which, as 'twould seem, 20
Is not without a reasonable ground:
But as physicians, when they seek to give
Young boys the nauseous wormwood, first do touch
The brim around the cup with the sweet juice
And yellow of the honey, in order that 25
The thoughtless age of boyhood be cajoled
As far as the lips, and meanwhile swallow down
The wormwood's bitter draught, and, though befooled,
Be yet not merely duped, but rather thus

Grow strong again with recreated health: 30
So now I too (since this my doctrine seems
In general somewhat woeful unto those
Who've had it not in hand, and since the crowd
Starts back from it in horror) have desired
To expound our doctrine unto thee in song 35
Soft-speaking and Pierian, and, as 'twere,
To touch it with sweet honey of the Muse–
If by such method haply I might hold
The mind of thee upon these lines of ours,
Till thou see through the nature of all things, 40
And how exists the interwoven frame.

But since I've taught that bodies of matter, made
Completely solid, hither and thither fly
Forevermore unconquered through all time, 45
Now come, and whether to the sum of them
There be a limit or be none, for thee
Let us unfold; likewise what has been found
To be the wide inane, or room, or space
Wherein all things soever do go on, 50
Let us examine if it finite be
All and entire, or reach unmeasured round
And downward an illimitable profound.

Thus, then, the All that is is limited 55
In no one region of its onward paths,
For then 'tmust have forever its beyond.
And a beyond 'tis seen can never be
For aught, unless still further on there be
A somewhat somewhere that may bound the same– 60
So that the thing be seen still on to where
The nature of sensation of that thing
Can follow it no longer. Now because
Confess we must there's naught beside the sum,
There's no beyond, and so it lacks all end. 65
It matters nothing where thou post thyself,
In whatsoever regions of the same;
Even any place a man has set him down
Still leaves about him the unbounded all
Outward in all directions; or, supposing 70

A moment the all of space finite to be,
If some one farthest traveller runs forth
Unto the extreme coasts and throws ahead
A flying spear, is't then thy wish to think
It goes, hurled off amain, to where 'twas sent 75
And shoots afar, or that some object there
Can thwart and stop it? For the one or other
Thou must admit and take. Either of which
Shuts off escape for thee, and does compel
That thou concede the all spreads everywhere, 80
Owning no confines. Since whether there be
Aught that may block and check it so it comes
Not where 'twas sent, nor lodges in its goal,
Or whether borne along, in either view
'Thas started not from any end. And so 85
I'll follow on, and whereso'er thou set
The extreme coasts, I'll query, "what becomes
Thereafter of thy spear?" 'Twill come to pass
That nowhere can a world's-end be, and that
The chance for further flight prolongs forever 90
The flight itself. Besides, were all the space
Of the totality and sum shut in
With fixed coasts, and bounded everywhere,
Then would the abundance of world's matter flow
Together by solid weight from everywhere 95
Still downward to the bottom of the world,
Nor aught could happen under cope of sky,
Nor could there be a sky at all or sun–
Indeed, where matter all one heap would lie,
By having settled during infinite time. 100
But in reality, repose is given
Unto no bodies 'mongst the elements,
Because there is no bottom whereunto
They might, as 'twere, together flow, and where
They might take up their undisturbed abodes. 105
In endless motion everything goes on
Forevermore; out of all regions, even
Out of the pit below, from forth the vast,
Are hurtled bodies evermore supplied.
The nature of room, the space of the abyss 110
Is such that even the flashing thunderbolts

Can neither speed upon their courses through,
Gliding across eternal tracts of time,
Nor, further, bring to pass, as on they run,
That they may bate their journeying one whit: 115
Such huge abundance spreads for things around–
Room off to every quarter, without end.
Lastly, before our very eyes is seen
Thing to bound thing: air hedges hill from hill,
And mountain walls hedge air; land ends the sea, 120
And sea in turn all lands; but for the All
Truly is nothing which outside may bound.
That, too, the sum of things itself may not
Have power to fix a measure of its own,
Great nature guards, she who compels the void 125
To bound all body, as body all the void,
Thus rendering by these alternates the whole
An infinite; or else the one or other,
Being unbounded by the other, spreads,
Even by its single nature, ne'ertheless 130
Immeasurably forth....
Nor sea, nor earth, nor shining vaults of sky,
Nor breed of mortals, nor holy limbs of gods
Could keep their place least portion of an hour:
For, driven apart from out its meetings fit, 135
The stock of stuff, dissolved, would be borne
Along the illimitable inane afar,
Or rather, in fact, would ne'er have once combined
And given a birth to aught, since, scattered wide,
It could not be united. For of truth 140
Neither by counsel did the primal germs
'Stablish themselves, as by keen act of mind,
Each in its proper place; nor did they make,
Forsooth, a compact how each germ should move;
But since, being many and changed in many modes 145
Along the All, they're driven abroad and vexed
By blow on blow, even from all time of old,
They thus at last, after attempting all
The kinds of motion and conjoining, come
Into those great arrangements out of which 150
This sum of things established is create,
By which, moreover, through the mighty years,

It is preserved, when once it has been thrown
Into the proper motions, bringing to pass
That ever the streams refresh the greedy main 155
With river-waves abounding, and that earth,
Lapped in warm exhalations of the sun,
Renews her broods, and that the lusty race
Of breathing creatures bears and blooms, and that
The gliding fires of ether are alive– 160
What still the primal germs nowise could do,
Unless from out the infinite of space
Could come supply of matter, whence in season
They're wont whatever losses to repair.
For as the nature of breathing creatures wastes, 165
Losing its body, when deprived of food:
So all things have to be dissolved as soon
As matter, diverted by what means soever
From off its course, shall fail to be on hand.
Nor can the blows from outward still conserve, 170
On every side, whatever sum of a world
Has been united in a whole. They can
Indeed, by frequent beating, check a part,
Till others arriving may fulfil the sum;
But meanwhile often are they forced to spring 175
Rebounding back, and, as they spring, to yield,
Unto those elements whence a world derives,
Room and a time for flight, permitting them
To be from off the massy union borne
Free and afar. Wherefore, again, again: 180
Needs must there come a many for supply;
And also, that the blows themselves shall be
Unfailing ever, must there ever be
An infinite force of matter all sides round.

185

And in these problems, shrink, my Memmius, far
From yielding faith to that notorious talk:
That all things inward to the centre press;
And thus the nature of the world stands firm
With never blows from outward, nor can be 190
Nowhere disparted–since all height and depth
Have always inward to the centre pressed
(If thou art ready to believe that aught

Itself can rest upon itself); or that
The ponderous bodies which be under earth 195
Do all press upwards and do come to rest
Upon the earth, in some way upside down,
Like to those images of things we see
At present through the waters. They contend,
With like procedure, that all breathing things 200
Head downward roam about, and yet cannot
Tumble from earth to realms of sky below,
No more than these our bodies wing away
Spontaneously to vaults of sky above;
That, when those creatures look upon the sun, 205
We view the constellations of the night;
And that with us the seasons of the sky
They thus alternately divide, and thus
Do pass the night coequal to our days,
But a vain error has given these dreams to fools, 210
Which they've embraced with reasoning perverse
For centre none can be where world is still
Boundless, nor yet, if now a centre were,
Could aught take there a fixed position more
Than for some other cause 'tmight be dislodged. 215
For all of room and space we call the void
Must both through centre and non-centre yield
Alike to weights where'er their motions tend.
Nor is there any place, where, when they've come,
Bodies can be at standstill in the void, 220
Deprived of force of weight; nor yet may void
Furnish support to any,–nay, it must,
True to its bent of nature, still give way.
Thus in such manner not at all can things
Be held in union, as if overcome 225
By craving for a centre.

But besides,
Seeing they feign that not all bodies press
To centre inward, rather only those 230
Of earth and water (liquid of the sea,
And the big billows from the mountain slopes,
And whatsoever are encased, as 'twere,
In earthen body), contrariwise, they teach

How the thin air, and with it the hot fire, 235
Is borne asunder from the centre, and how,
For this all ether quivers with bright stars,
And the sun's flame along the blue is fed
(Because the heat, from out the centre flying,
All gathers there), and how, again, the boughs 240
Upon the tree-tops could not sprout their leaves,
Unless, little by little, from out the earth
For each were nutriment...

Lest, after the manner of the winged flames, 245
The ramparts of the world should flee away,
Dissolved amain throughout the mighty void,
And lest all else should likewise follow after,
Aye, lest the thundering vaults of heaven should burst
And splinter upward, and the earth forthwith 250
Withdraw from under our feet, and all its bulk,
Among its mingled wrecks and those of heaven,
With slipping asunder of the primal seeds,
Should pass, along the immeasurable inane,
Away forever, and, that instant, naught 255
Of wrack and remnant would be left, beside
The desolate space, and germs invisible.
For on whatever side thou deemest first
The primal bodies lacking, lo, that side
Will be for things the very door of death: 260
Wherethrough the throng of matter all will dash,
Out and abroad.

These points, if thou wilt ponder,
Then, with but paltry trouble led along... 265
For one thing after other will grow clear,
Nor shall the blind night rob thee of the road,
To hinder thy gaze on nature's Farthest-forth.
Thus things for things shall kindle torches new.

Book Two

Proem

'Tis sweet, when, down the mighty main, the winds
Roll up its waste of waters, from the land
To watch another's labouring anguish far,
Not that we joyously delight that man
Should thus be smitten, but because 'tis sweet 5
To mark what evils we ourselves be spared;
'Tis sweet, again, to view the mighty strife
Of armies embattled yonder o'er the plains,
Ourselves no sharers in the peril; but naught
There is more goodly than to hold the high 10
Serene plateaus, well fortressed by the wise,
Whence thou may'st look below on other men
And see them ev'rywhere wand'ring, all dispersed
In their lone seeking for the road of life;
Rivals in genius, or emulous in rank, 15
Pressing through days and nights with hugest toil
For summits of power and mastery of the world.
O wretched minds of men! O blinded hearts!
In how great perils, in what darks of life
Are spent the human years, however brief!– 20
O not to see that nature for herself
Barks after nothing, save that pain keep off,
Disjoined from the body, and that mind enjoy
Delightsome feeling, far from care and fear!
Therefore we see that our corporeal life 25
Needs little, altogether, and only such

As takes the pain away, and can besides
Strew underneath some number of delights.
More grateful 'tis at times (for nature craves
No artifice nor luxury), if forsooth 30
There be no golden images of boys
Along the halls, with right hands holding out
The lamps ablaze, the lights for evening feasts,
And if the house doth glitter not with gold
Nor gleam with silver, and to the lyre resound 35
No fretted and gilded ceilings overhead,
Yet still to lounge with friends in the soft grass
Beside a river of water, underneath
A big tree's boughs, and merrily to refresh
Our frames, with no vast outlay–most of all 40
If the weather is laughing and the times of the year
Besprinkle the green of the grass around with flowers.
Nor yet the quicker will hot fevers go,
If on a pictured tapestry thou toss,
Or purple robe, than if 'tis thine to lie 45
Upon the poor man's bedding. Wherefore, since
Treasure, nor rank, nor glory of a reign
Avail us naught for this our body, thus
Reckon them likewise nothing for the mind:
Save then perchance, when thou beholdest forth 50
Thy legions swarming round the Field of Mars,
Rousing a mimic warfare–either side
Strengthened with large auxiliaries and horse,
Alike equipped with arms, alike inspired;
Or save when also thou beholdest forth 55
Thy fleets to swarm, deploying down the sea:
For then, by such bright circumstance abashed,
Religion pales and flees thy mind; O then
The fears of death leave heart so free of care.
But if we note how all this pomp at last 60
Is but a drollery and a mocking sport,
And of a truth man's dread, with cares at heels,
Dreads not these sounds of arms, these savage swords
But among kings and lords of all the world
Mingles undaunted, nor is overawed 65
By gleam of gold nor by the splendour bright
Of purple robe, canst thou then doubt that this

Is aught, but power of thinking?–when, besides
The whole of life but labours in the dark.
For just as children tremble and fear all 70
In the viewless dark, so even we at times
Dread in the light so many things that be
No whit more fearsome than what children feign,
Shuddering, will be upon them in the dark.
This terror then, this darkness of the mind, 75
Not sunrise with its flaring spokes of light,
Nor glittering arrows of morning can disperse,
But only nature's aspect and her law.

Atomic Motions

Now come: I will untangle for thy steps
Now by what motions the begetting bodies
Of the world-stuff beget the varied world,
And then forever resolve it when begot,
And by what force they are constrained to this, 5
And what the speed appointed unto them
Wherewith to travel down the vast inane:
Do thou remember to yield thee to my words.
For truly matter coheres not, crowds not tight,
Since we behold each thing to wane away, 10
And we observe how all flows on and off,
As 'twere, with age-old time, and from our eyes
How eld withdraws each object at the end,
Albeit the sum is seen to bide the same,
Unharmed, because these motes that leave each thing 15
Diminish what they part from, but endow
With increase those to which in turn they come,
Constraining these to wither in old age,
And those to flower at the prime (and yet
Biding not long among them). Thus the sum 20
Forever is replenished, and we live
As mortals by eternal give and take.
The nations wax, the nations wane away;
In a brief space the generations pass,
And like to runners hand the lamp of life 25
One unto other.

But if thou believe
That the primordial germs of things can stop,

And in their stopping give new motions birth, 30
Afar thou wanderest from the road of truth.
For since they wander through the void inane,
All the primordial germs of things must needs
Be borne along, either by weight their own,
Or haply by another's blow without. 35
For, when, in their incessancy so oft
They meet and clash, it comes to pass amain
They leap asunder, face to face: not strange–
Being most hard, and solid in their weights,
And naught opposing motion, from behind. 40
And that more clearly thou perceive how all
These mites of matter are darted round about,
Recall to mind how nowhere in the sum
Of All exists a bottom,–nowhere is
A realm of rest for primal bodies; since 45
(As amply shown and proved by reason sure)
Space has no bound nor measure, and extends
Unmetered forth in all directions round.
Since this stands certain, thus 'tis out of doubt
No rest is rendered to the primal bodies 50
Along the unfathomable inane; but rather,
Inveterately plied by motions mixed,
Some, at their jamming, bound aback and leave
Huge gaps between, and some from off the blow
Are hurried about with spaces small between. 55
And all which, brought together with slight gaps,
In more condensed union bound aback,
Linked by their own all inter-tangled shapes,–
These form the irrefragable roots of rocks
And the brute bulks of iron, and what else 60
Is of their kind...
The rest leap far asunder, far recoil,
Leaving huge gaps between: and these supply
For us thin air and splendour-lights of the sun.
And many besides wander the mighty void– 65
Cast back from unions of existing things,
Nowhere accepted in the universe,
And nowise linked in motions to the rest.
And of this fact (as I record it here)
An image, a type goes on before our eyes 70

Present each moment; for behold whenever
The sun's light and the rays, let in, pour down
Across dark halls of houses: thou wilt see
The many mites in many a manner mixed
Amid a void in the very light of the rays, 75
And battling on, as in eternal strife,
And in battalions contending without halt,
In meetings, partings, harried up and down.
From this thou mayest conjecture of what sort
The ceaseless tossing of primordial seeds 80
Amid the mightier void–at least so far
As small affair can for a vaster serve,
And by example put thee on the spoor
Of knowledge. For this reason too 'tis fit
Thou turn thy mind the more unto these bodies 85
Which here are witnessed tumbling in the light:
Namely, because such tumblings are a sign
That motions also of the primal stuff
Secret and viewless lurk beneath, behind.
For thou wilt mark here many a speck, impelled 90
By viewless blows, to change its little course,
And beaten backwards to return again,
Hither and thither in all directions round.
Lo, all their shifting movement is of old,
From the primeval atoms; for the same 95
Primordial seeds of things first move of self,
And then those bodies built of unions small
And nearest, as it were, unto the powers
Of the primeval atoms, are stirred up
By impulse of those atoms' unseen blows, 100
And these thereafter goad the next in size:
Thus motion ascends from the primevals on,
And stage by stage emerges to our sense,
Until those objects also move which we
Can mark in sunbeams, though it not appears 105
What blows do urge them.

Herein wonder not
How 'tis that, while the seeds of things are all
Moving forever, the sum yet seems to stand 110
Supremely still, except in cases where

A thing shows motion of its frame as whole.
For far beneath the ken of senses lies
The nature of those ultimates of the world;
And so, since those themselves thou canst not see, 115
Their motion also must they veil from men–
For mark, indeed, how things we can see, oft
Yet hide their motions, when afar from us
Along the distant landscape. Often thus,
Upon a hillside will the woolly flocks 120
Be cropping their goodly food and creeping about
Whither the summons of the grass, begemmed
With the fresh dew, is calling, and the lambs,
Well filled, are frisking, locking horns in sport:
Yet all for us seem blurred and blent afar– 125
A glint of white at rest on a green hill.
Again, when mighty legions, marching round,
Fill all the quarters of the plains below,
Rousing a mimic warfare, there the sheen
Shoots up the sky, and all the fields about 130
Glitter with brass, and from beneath, a sound
Goes forth from feet of stalwart soldiery,
And mountain walls, smote by the shouting, send
The voices onward to the stars of heaven,
And hither and thither darts the cavalry, 135
And of a sudden down the midmost fields
Charges with onset stout enough to rock
The solid earth: and yet some post there is
Up the high mountains, viewed from which they seem
To stand–a gleam at rest along the plains. 140

Now what the speed to matter's atoms given
Thou mayest in few, my Memmius, learn from this:
When first the dawn is sprinkling with new light
The lands, and all the breed of birds abroad 145
Flit round the trackless forests, with liquid notes
Filling the regions along the mellow air,
We see 'tis forthwith manifest to man
How suddenly the risen sun is wont
At such an hour to overspread and clothe 150
The whole with its own splendour; but the sun's
Warm exhalations and this serene light

Travel not down an empty void; and thus
They are compelled more slowly to advance,
Whilst, as it were, they cleave the waves of air; 155
Nor one by one travel these particles
Of the warm exhalations, but are all
Entangled and enmassed, whereby at once
Each is restrained by each, and from without
Checked, till compelled more slowly to advance. 160
But the primordial atoms with their old
Simple solidity, when forth they travel
Along the empty void, all undelayed
By aught outside them there, and they, each one
Being one unit from nature of its parts, 165
Are borne to that one place on which they strive
Still to lay hold, must then, beyond a doubt,
Outstrip in speed, and be more swiftly borne
Than light of sun, and over regions rush,
Of space much vaster, in the self-same time 170
The sun's effulgence widens round the sky.

Nor to pursue the atoms one by one,
To see the law whereby each thing goes on.
But some men, ignorant of matter, think, 175
Opposing this, that not without the gods,
In such adjustment to our human ways,
Can nature change the seasons of the years,
And bring to birth the grains and all of else
To which divine Delight, the guide of life, 180
Persuades mortality and leads it on,
That, through her artful blandishments of love,
It propagate the generations still,
Lest humankind should perish. When they feign
That gods have stablished all things but for man, 185
They seem in all ways mightily to lapse
From reason's truth: for ev'n if ne'er I knew
What seeds primordial are, yet would I dare
This to affirm, ev'n from deep judgment based
Upon the ways and conduct of the skies– 190
This to maintain by many a fact besides–
That in no wise the nature of the world
For us was builded by a power divine–

So great the faults it stands encumbered with:
The which, my Memmius, later on, for thee 195
We will clear up. Now as to what remains
Concerning motions we'll unfold our thought.

Now is the place, meseems, in these affairs
To prove for thee this too: nothing corporeal 200
Of its own force can e'er be upward borne,
Or upward go–nor let the bodies of flames
Deceive thee here: for they engendered are
With urge to upwards, taking thus increase,
Whereby grow upwards shining grains and trees, 205
Though all the weight within them downward bears.
Nor, when the fires will leap from under round
The roofs of houses, and swift flame laps up
Timber and beam, 'tis then to be supposed
They act of own accord, no force beneath 210
To urge them up. 'Tis thus that blood, discharged
From out our bodies, spurts its jets aloft
And spatters gore. And hast thou never marked
With what a force the water will disgorge
Timber and beam? The deeper, straight and down, 215
We push them in, and, many though we be,
The more we press with main and toil, the more
The water vomits up and flings them back,
That, more than half their length, they there emerge,
Rebounding. Yet we never doubt, meseems, 220
That all the weight within them downward bears
Through empty void. Well, in like manner, flames
Ought also to be able, when pressed out,
Through winds of air to rise aloft, even though
The weight within them strive to draw them down. 225
Hast thou not seen, sweeping so far and high,
The meteors, midnight flambeaus of the sky,
How after them they draw long trails of flame
Wherever Nature gives a thoroughfare?
How stars and constellations drop to earth, 230
Seest not? Nay, too, the sun from peak of heaven
Sheds round to every quarter its large heat,
And sows the new-ploughed intervales with light:
Thus also sun's heat downward tends to earth.

Athwart the rain thou seest the lightning fly; 235
Now here, now there, bursting from out the clouds,
The fires dash zig-zag–and that flaming power
Falls likewise down to earth.

In these affairs 240
We wish thee also well aware of this:
The atoms, as their own weight bears them down
Plumb through the void, at scarce determined times,
In scarce determined places, from their course
Decline a little–call it, so to speak, 245
Mere changed trend. For were it not their wont
Thuswise to swerve, down would they fall, each one,
Like drops of rain, through the unbottomed void;
And then collisions ne'er could be nor blows
Among the primal elements; and thus 250
Nature would never have created aught.

But, if perchance be any that believe
The heavier bodies, as more swiftly borne
Plumb down the void, are able from above 255
To strike the lighter, thus engendering blows
Able to cause those procreant motions, far
From highways of true reason they retire.
For whatsoever through the waters fall,
Or through thin air, must quicken their descent, 260
Each after its weight–on this account, because
Both bulk of water and the subtle air
By no means can retard each thing alike,
But give more quick before the heavier weight;
But contrariwise the empty void cannot, 265
On any side, at any time, to aught
Oppose resistance, but will ever yield,
True to its bent of nature. Wherefore all,
With equal speed, though equal not in weight,
Must rush, borne downward through the still inane. 270
Thus ne'er at all have heavier from above
Been swift to strike the lighter, gendering strokes
Which cause those divers motions, by whose means
Nature transacts her work. And so I say,
The atoms must a little swerve at times– 275

But only the least, lest we should seem to feign
Motions oblique, and fact refute us there.
For this we see forthwith is manifest:
Whatever the weight, it can't obliquely go,
Down on its headlong journey from above, 280
At least so far as thou canst mark; but who
Is there can mark by sense that naught can swerve
At all aside from off its road's straight line?

Again, if ev'r all motions are co-linked, 285
And from the old ever arise the new
In fixed order, and primordial seeds
Produce not by their swerving some new start
Of motion to sunder the covenants of fate,
That cause succeed not cause from everlasting, 290
Whence this free will for creatures o'er the lands,
Whence is it wrested from the fates,–this will
Whereby we step right forward where desire
Leads each man on, whereby the same we swerve
In motions, not as at some fixed time, 295
Nor at some fixed line of space, but where
The mind itself has urged? For out of doubt
In these affairs 'tis each man's will itself
That gives the start, and hence throughout our limbs
Incipient motions are diffused. Again, 300
Dost thou not see, when, at a point of time,
The bars are opened, how the eager strength
Of horses cannot forward break as soon
As pants their mind to do? For it behooves
That all the stock of matter, through the frame, 305
Be roused, in order that, through every joint,
Aroused, it press and follow mind's desire;
So thus thou seest initial motion's gendered
From out the heart, aye, verily, proceeds
First from the spirit's will, whence at the last 310
'Tis given forth through joints and body entire.
Quite otherwise it is, when forth we move,
Impelled by a blow of another's mighty powers
And mighty urge; for then 'tis clear enough
All matter of our total body goes, 315
Hurried along, against our own desire–

Until the will has pulled upon the reins
And checked it back, throughout our members all;
At whose arbitrament indeed sometimes
The stock of matter's forced to change its path, 320
Throughout our members and throughout our joints,
And, after being forward cast, to be
Reined up, whereat it settles back again.
So seest thou not, how, though external force
Drive men before, and often make them move, 325
Onward against desire, and headlong snatched,
Yet is there something in these breasts of ours
Strong to combat, strong to withstand the same?–
Wherefore no less within the primal seeds
Thou must admit, besides all blows and weight, 330
Some other cause of motion, whence derives
This power in us inborn, of some free act.–
Since naught from nothing can become, we see.
For weight prevents all things should come to pass
Through blows, as 'twere, by some external force; 335
But that man's mind itself in all it does
Hath not a fixed necessity within,
Nor is not, like a conquered thing, compelled
To bear and suffer,–this state comes to man
From that slight swervement of the elements 340
In no fixed line of space, in no fixed time.

Nor ever was the stock of stuff more crammed,
Nor ever, again, sundered by bigger gaps:
For naught gives increase and naught takes away; 345
On which account, just as they move to-day,
The elemental bodies moved of old
And shall the same hereafter evermore.
And what was wont to be begot of old
Shall be begotten under selfsame terms 350
And grow and thrive in power, so far as given
To each by Nature's changeless, old decrees.
The sum of things there is no power can change,
For naught exists outside, to which can flee
Out of the world matter of any kind, 355
Nor forth from which a fresh supply can spring,
Break in upon the founded world, and change
Whole nature of things, and turn their motions about.

Atomic Forms and Their Combinations

Now come, and next hereafter apprehend
What sorts, how vastly different in form,
How varied in multitudinous shapes they are–
These old beginnings of the universe;
Not in the sense that only few are furnished 5
With one like form, but rather not at all
In general have they likeness each with each,
No marvel: since the stock of them's so great
That there's no end (as I have taught) nor sum,
They must indeed not one and all be marked 10
By equal outline and by shape the same.

Moreover, humankind, and the mute flocks
Of scaly creatures swimming in the streams,
And joyous herds around, and all the wild, 15
And all the breeds of birds–both those that teem
In gladsome regions of the water-haunts,
About the river-banks and springs and pools,
And those that throng, flitting from tree to tree,
Through trackless woods–Go, take which one thou wilt, 20
In any kind: thou wilt discover still
Each from the other still unlike in shape.
Nor in no other wise could offspring know
Mother, nor mother offspring–which we see
They yet can do, distinguished one from other, 25
No less than human beings, by clear signs.
Thus oft before fair temples of the gods,

Beside the incense-burning altars slain,
Drops down the yearling calf, from out its breast
Breathing warm streams of blood; the orphaned mother, 30
Ranging meanwhile green woodland pastures round,
Knows well the footprints, pressed by cloven hoofs,
With eyes regarding every spot about,
For sight somewhere of youngling gone from her;
And, stopping short, filleth the leafy lanes 35
With her complaints; and oft she seeks again
Within the stall, pierced by her yearning still.
Nor tender willows, nor dew-quickened grass,
Nor the loved streams that glide along low banks,
Can lure her mind and turn the sudden pain; 40
Nor other shapes of calves that graze thereby
Distract her mind or lighten pain the least–
So keen her search for something known and hers.
Moreover, tender kids with bleating throats
Do know their horned dams, and butting lambs 45
The flocks of sheep, and thus they patter on,
Unfailingly each to its proper teat,
As nature intends. Lastly, with any grain,
Thou'lt see that no one kernel in one kind
Is so far like another, that there still 50
Is not in shapes some difference running through.
By a like law we see how earth is pied
With shells and conchs, where, with soft waves, the sea
Beats on the thirsty sands of curving shores.
Wherefore again, again, since seeds of things 55
Exist by nature, nor were wrought with hands
After a fixed pattern of one other,
They needs must flitter to and fro with shapes
In types dissimilar to one another.

60

Easy enough by thought of mind to solve
Why fires of lightning more can penetrate
Than these of ours from pitch-pine born on earth.
For thou canst say lightning's celestial fire,
So subtle, is formed of figures finer far, 65
And passes thus through holes which this our fire,
Born from the wood, created from the pine,
Cannot. Again, light passes through the horn

On the lantern's side, while rain is dashed away.
And why?–unless those bodies of light should be 70
Finer than those of water's genial showers.
We see how quickly through a colander
The wines will flow; how, on the other hand,
The sluggish olive-oil delays: no doubt,
Because 'tis wrought of elements more large, 75
Or else more crook'd and intertangled. Thus
It comes that the primordials cannot be
So suddenly sundered one from other, and seep,
One through each several hole of anything.
80

And note, besides, that liquor of honey or milk
Yields in the mouth agreeable taste to tongue,
Whilst nauseous wormwood, pungent centaury,
With their foul flavour set the lips awry;
Thus simple 'tis to see that whatsoever 85
Can touch the senses pleasingly are made
Of smooth and rounded elements, whilst those
Which seem the bitter and the sharp, are held
Entwined by elements more crook'd, and so
Are wont to tear their ways into our senses, 90
And rend our body as they enter in.
In short all good to sense, all bad to touch,
Being up-built of figures so unlike,
Are mutually at strife–lest thou suppose
That the shrill rasping of a squeaking saw 95
Consists of elements as smooth as song
Which, waked by nimble fingers, on the strings
The sweet musicians fashion; or suppose
That same-shaped atoms through men's nostrils pierce
When foul cadavers burn, as when the stage 100
Is with Cilician saffron sprinkled fresh,
And the altar near exhales Panchaean scent;
Or hold as of like seed the goodly hues
Of things which feast our eyes, as those which sting
Against the smarting pupil and draw tears, 105
Or show, with gruesome aspect, grim and vile.
For never a shape which charms our sense was made
Without some elemental smoothness; whilst
Whate'er is harsh and irksome has been framed

Still with some roughness in its elements. 110
Some, too, there are which justly are supposed
To be nor smooth nor altogether hooked,
With bended barbs, but slightly angled-out,
To tickle rather than to wound the sense–
And of which sort is the salt tartar of wine 115
And flavours of the gummed elecampane.
Again, that glowing fire and icy rime
Are fanged with teeth unlike whereby to sting
Our body's sense, the touch of each gives proof.
For touch–by sacred majesties of Gods!– 120
Touch is indeed the body's only sense–
Be't that something in-from-outward works,
Be't that something in the body born
Wounds, or delighteth as it passes out
Along the procreant paths of Aphrodite; 125
Or be't the seeds by some collision whirl
Disordered in the body and confound
By tumult and confusion all the sense–
As thou mayst find, if haply with the hand
Thyself thou strike thy body's any part. 130
On which account, the elemental forms
Must differ widely, as enabled thus
To cause diverse sensations.

And, again, 135
What seems to us the hardened and condensed
Must be of atoms among themselves more hooked,
Be held compacted deep within, as 'twere
By branch-like atoms–of which sort the chief
Are diamond stones, despisers of all blows, 140
And stalwart flint and strength of solid iron,
And brazen bars, which, budging hard in locks,
Do grate and scream. But what are liquid, formed
Of fluid body, they indeed must be
Of elements more smooth and round–because 145
Their globules severally will not cohere:
To suck the poppy-seeds from palm of hand
Is quite as easy as drinking water down,
And they, once struck, roll like unto the same.
But that thou seest among the things that flow 150

Some bitter, as the brine of ocean is,
Is not the least a marvel...
For since 'tis fluid, smooth its atoms are
And round, with painful rough ones mixed therein;
Yet need not these be held together hooked: 155
In fact, though rough, they're globular besides,
Able at once to roll, and rasp the sense.
And that the more thou mayst believe me here,
That with smooth elements are mixed the rough
(Whence Neptune's salt astringent body comes), 160
There is a means to separate the twain,
And thereupon dividedly to see
How the sweet water, after filtering through
So often underground, flows freshened forth
Into some hollow; for it leaves above 165
The primal germs of nauseating brine,
Since cling the rough more readily in earth.
Lastly, whatso thou markest to disperse
Upon the instant–smoke, and cloud, and flame–
Must not (even though not all of smooth and round) 170
Be yet co-linked with atoms intertwined,
That thus they can, without together cleaving,
So pierce our body and so bore the rocks.
Whatever we see...
Given to senses, that thou must perceive 175
They're not from linked but pointed elements.

The which now having taught, I will go on
To bind thereto a fact to this allied
And drawing from this its proof: these primal germs 180
Vary, yet only with finite tale of shapes.
For were these shapes quite infinite, some seeds
Would have a body of infinite increase.
For in one seed, in one small frame of any,
The shapes can't vary from one another much. 185
Assume, we'll say, that of three minim parts
Consist the primal bodies, or add a few:
When, now, by placing all these parts of one
At top and bottom, changing lefts and rights,
Thou hast with every kind of shift found out 190
What the aspect of shape of its whole body

Each new arrangement gives, for what remains,
If thou percase wouldst vary its old shapes,
New parts must then be added; follows next,
If thou percase wouldst vary still its shapes, 195
That by like logic each arrangement still
Requires its increment of other parts.
Ergo, an augmentation of its frame
Follows upon each novelty of forms.
Wherefore, it cannot be thou'lt undertake 200
That seeds have infinite differences in form,
Lest thus thou forcest some indeed to be
Of an immeasurable immensity–
Which I have taught above cannot be proved.

205

And now for thee barbaric robes, and gleam
Of Meliboean purple, touched with dye
Of the Thessalian shell...
The peacock's golden generations, stained
With spotted gaieties, would lie o'erthrown 210
By some new colour of new things more bright;
The odour of myrrh and savours of honey despised;
The swan's old lyric, and Apollo's hymns,
Once modulated on the many chords,
Would likewise sink o'ermastered and be mute: 215
For, lo, a somewhat, finer than the rest,
Would be arising evermore. So, too,
Into some baser part might all retire,
Even as we said to better might they come:
For, lo, a somewhat, loathlier than the rest 220
To nostrils, ears, and eyes, and taste of tongue,
Would then, by reasoning reversed, be there.
Since 'tis not so, but unto things are given
Their fixed limitations which do bound
Their sum on either side, 'tmust be confessed 225
That matter, too, by finite tale of shapes
Does differ. Again, from earth's midsummer heats
Unto the icy hoar-frosts of the year
The forward path is fixed, and by like law
O'ertravelled backwards at the dawn of spring. 230
For each degree of hot, and each of cold,
And the half-warm, all filling up the sum

In due progression, lie, my Memmius, there
Betwixt the two extremes: the things create
Must differ, therefore, by a finite change, 235
Since at each end marked off they ever are
By fixed point–on one side plagued by flames
And on the other by congealing frosts.

The which now having taught, I will go on 240
To bind thereto a fact to this allied
And drawing from this its proof: those primal germs
Which have been fashioned all of one like shape
Are infinite in tale; for, since the forms
Themselves are finite in divergences, 245
Then those which are alike will have to be
Infinite, else the sum of stuff remains
A finite–what I've proved is not the fact,
Showing in verse how corpuscles of stuff,
From everlasting and to-day the same, 250
Uphold the sum of things, all sides around
By old succession of unending blows.
For though thou view'st some beasts to be more rare,
And mark'st in them a less prolific stock,
Yet in another region, in lands remote, 255
That kind abounding may make up the count;
Even as we mark among the four-foot kind
Snake-handed elephants, whose thousands wall
With ivory ramparts India about,
That her interiors cannot entered be– 260
So big her count of brutes of which we see
Such few examples. Or suppose, besides,
We feign some thing, one of its kind and sole
With body born, to which is nothing like
In all the lands: yet now unless shall be 265
An infinite count of matter out of which
Thus to conceive and bring it forth to life,
It cannot be created and–what's more–
It cannot take its food and get increase.
Yea, if through all the world in finite tale 270
Be tossed the procreant bodies of one thing,
Whence, then, and where in what mode, by what power,
Shall they to meeting come together there,

In such vast ocean of matter and tumult strange?–
No means they have of joining into one. 275
But, just as, after mighty ship-wrecks piled,
The mighty main is wont to scatter wide
The rowers' banks, the ribs, the yards, the prow,
The masts and swimming oars, so that afar
Along all shores of lands are seen afloat 280
The carven fragments of the rended poop,
Giving a lesson to mortality
To shun the ambush of the faithless main,
The violence and the guile, and trust it not
At any hour, however much may smile 285
The crafty enticements of the placid deep:
Exactly thus, if once thou holdest true
That certain seeds are finite in their tale,
The various tides of matter, then, must needs
Scatter them flung throughout the ages all, 290
So that not ever can they join, as driven
Together into union, nor remain
In union, nor with increment can grow–
But facts in proof are manifest for each:
Things can be both begotten and increase. 295
'Tis therefore manifest that primal germs,
Are infinite in any class thou wilt–
From whence is furnished matter for all things.

Nor can those motions that bring death prevail 300
Forever, nor eternally entomb
The welfare of the world; nor, further, can
Those motions that give birth to things and growth
Keep them forever when created there.
Thus the long war, from everlasting waged, 305
With equal strife among the elements
Goes on and on. Now here, now there, prevail
The vital forces of the world–or fall.
Mixed with the funeral is the wildered wail
Of infants coming to the shores of light: 310
No night a day, no dawn a night hath followed
That heard not, mingling with the small birth-cries,
The wild laments, companions old of death
And the black rites.

This, too, in these affairs 315
'Tis fit thou hold well sealed, and keep consigned
With no forgetting brain: nothing there is
Whose nature is apparent out of hand
That of one kind of elements consists–
Nothing there is that's not of mixed seed. 320
And whatsoe'er possesses in itself
More largely many powers and properties
Shows thus that here within itself there are
The largest number of kinds and differing shapes
Of elements. And, chief of all, the earth 325
Hath in herself first bodies whence the springs,
Rolling chill waters, renew forevermore
The unmeasured main; hath whence the fires arise–
For burns in many a spot her flamed crust,
Whilst the impetuous Aetna raves indeed 330
From more profounder fires–and she, again,
Hath in herself the seed whence she can raise
The shining grains and gladsome trees for men;
Whence, also, rivers, fronds, and gladsome pastures
Can she supply for mountain-roaming beasts. 335
Wherefore great mother of gods, and mother of beasts,
And parent of man hath she alone been named.

Her hymned the old and learned bards of Greece

340

Seated in chariot o'er the realms of air
To drive her team of lions, teaching thus
That the great earth hangs poised and cannot lie
Resting on other earth. Unto her car
They've yoked the wild beasts, since a progeny, 345
However savage, must be tamed and chid
By care of parents. They have girt about
With turret-crown the summit of her head,
Since, fortressed in her goodly strongholds high,
'Tis she sustains the cities; now, adorned 350
With that same token, to-day is carried forth,
With solemn awe through many a mighty land,
The image of that mother, the divine.
Her the wide nations, after antique rite,
Do name Idaean Mother, giving her 355

Escort of Phrygian bands, since first, they say,
From out those regions 'twas that grain began
Through all the world. To her do they assign
The Galli, the emasculate, since thus
They wish to show that men who violate 360
The majesty of the mother and have proved
Ingrate to parents are to be adjudged
Unfit to give unto the shores of light
A living progeny. The Galli come:
And hollow cymbals, tight-skinned tambourines 365
Resound around to bangings of their hands;
The fierce horns threaten with a raucous bray;
The tubed pipe excites their maddened minds
In Phrygian measures; they bear before them knives,
Wild emblems of their frenzy, which have power 370
The rabble's ingrate heads and impious hearts
To panic with terror of the goddess' might.
And so, when through the mighty cities borne,
She blesses man with salutations mute,
They strew the highway of her journeyings 375
With coin of brass and silver, gifting her
With alms and largesse, and shower her and shade
With flowers of roses falling like the snow
Upon the Mother and her companion-bands.
Here is an armed troop, the which by Greeks 380
Are called the Phrygian Curetes. Since
Haply among themselves they use to play
In games of arms and leap in measure round
With bloody mirth and by their nodding shake
The terrorizing crests upon their heads, 385
This is the armed troop that represents
The arm'd Dictaean Curetes, who, in Crete,
As runs the story, whilom did out-drown
That infant cry of Zeus, what time their band,
Young boys, in a swift dance around the boy, 390
To measured step beat with the brass on brass,
That Saturn might not get him for his jaws,
And give its mother an eternal wound
Along her heart. And 'tis on this account
That armed they escort the mighty Mother, 395
Or else because they signify by this

That she, the goddess, teaches men to be
Eager with armed valour to defend
Their motherland, and ready to stand forth,
The guard and glory of their parents' years. 400
A tale, however beautifully wrought,
That's wide of reason by a long remove:
For all the gods must of themselves enjoy
Immortal aeons and supreme repose,
Withdrawn from our affairs, detached, afar: 405
Immune from peril and immune from pain,
Themselves abounding in riches of their own,
Needing not us, they are not touched by wrath
They are not taken by service or by gift.
Truly is earth insensate for all time; 410
But, by obtaining germs of many things,
In many a way she brings the many forth
Into the light of sun. And here, whoso
Decides to call the ocean Neptune, or
The grain-crop Ceres, and prefers to abuse 415
The name of Bacchus rather than pronounce
The liquor's proper designation, him
Let us permit to go on calling earth
Mother of Gods, if only he will spare
To taint his soul with foul religion. 420
So, too, the wooly flocks, and horned kine,
And brood of battle-eager horses, grazing
Often together along one grassy plain,
Under the cope of one blue sky, and slaking
From out one stream of water each its thirst, 425
All live their lives with face and form unlike,
Keeping the parents' nature, parents' habits,
Which, kind by kind, through ages they repeat.
So great in any sort of herb thou wilt,
So great again in any river of earth 430
Are the distinct diversities of matter.
Hence, further, every creature–any one
From out them all–compounded is the same
Of bones, blood, veins, heat, moisture, flesh, and thews–
All differing vastly in their forms, and built 435
Of elements dissimilar in shape.
Again, all things by fire consumed ablaze,

Within their frame lay up, if naught besides,
At least those atoms whence derives their power
To throw forth fire and send out light from under, 440
To shoot the sparks and scatter embers wide.
If, with like reasoning of mind, all else
Thou traverse through, thou wilt discover thus
That in their frame the seeds of many things
They hide, and divers shapes of seeds contain. 445
Further, thou markest much, to which are given
Along together colour and flavour and smell,
Among which, chief, are most burnt offerings.

Thus must they be of divers shapes composed. 450
A smell of scorching enters in our frame
Where the bright colour from the dye goes not;
And colour in one way, flavour in quite another
Works inward to our senses–so mayst see
They differ too in elemental shapes. 455
Thus unlike forms into one mass combine,
And things exist by intermixed seed.

But still 'tmust not be thought that in all ways
All things can be conjoined; for then wouldst view 460
Portents begot about thee every side:
Hulks of mankind half brute astarting up,
At times big branches sprouting from man's trunk,
Limbs of a sea-beast to a land-beast knit,
And nature along the all-producing earth 465
Feeding those dire Chimaeras breathing flame
From hideous jaws–Of which 'tis simple fact
That none have been begot; because we see
All are from fixed seed and fixed dam
Engendered and so function as to keep 470
Throughout their growth their own ancestral type.
This happens surely by a fixed law:
For from all food-stuff, when once eaten down,
Go sundered atoms, suited to each creature,
Throughout their bodies, and, conjoining there, 475
Produce the proper motions; but we see
How, contrariwise, nature upon the ground
Throws off those foreign to their frame; and many

With viewless bodies from their bodies fly,
By blows impelled–those impotent to join 480
To any part, or, when inside, to accord
And to take on the vital motions there.
But think not, haply, living forms alone
Are bound by these laws: they distinguished all.

485

For just as all things of creation are,
In their whole nature, each to each unlike,
So must their atoms be in shape unlike–
Not since few only are fashioned of like form,
But since they all, as general rule, are not 490
The same as all. Nay, here in these our verses,
Elements many, common to many words,
Thou seest, though yet 'tis needful to confess
The words and verses differ, each from each,
Compounded out of different elements– 495
Not since few only, as common letters, run
Through all the words, or no two words are made,
One and the other, from all like elements,
But since they all, as general rule, are not
The same as all. Thus, too, in other things, 500
Whilst many germs common to many things
There are, yet they, combined among themselves,
Can form new wholes to others quite unlike.
Thus fairly one may say that humankind,
The grains, the gladsome trees, are all made up 505
Of different atoms. Further, since the seeds
Are different, difference must there also be
In intervening spaces, thoroughfares,
Connections, weights, blows, clashings, motions, all
Which not alone distinguish living forms, 510
But sunder earth's whole ocean from the lands,
And hold all heaven from the lands away.

ABSENCE OF SECONDARY QUALITIES

Now come, this wisdom by my sweet toil sought 515
Look thou perceive, lest haply thou shouldst guess
That the white objects shining to thine eyes
Are gendered of white atoms, or the black

Of a black seed; or yet believe that aught
That's steeped in any hue should take its dye 520
From bits of matter tinct with hue the same.
For matter's bodies own no hue the least–
Or like to objects or, again, unlike.
But, if percase it seem to thee that mind
Itself can dart no influence of its own 525
Into these bodies, wide thou wand'rest off.
For since the blind-born, who have ne'er surveyed
The light of sun, yet recognise by touch
Things that from birth had ne'er a hue for them,
'Tis thine to know that bodies can be brought 530
No less unto the ken of our minds too,
Though yet those bodies with no dye be smeared.
Again, ourselves whatever in the dark
We touch, the same we do not find to be
Tinctured with any colour. 535

Now that here
I win the argument, I next will teach

Now, every colour changes, none except, 540
And every...
Which the primordials ought nowise to do.
Since an immutable somewhat must remain,
Lest all things utterly be brought to naught.
For change of anything from out its bounds 545
Means instant death of that which was before.
Wherefore be mindful not to stain with colour
The seeds of things, lest things return for thee
All utterly to naught.

550

But now, if seeds
Receive no property of colour, and yet
Be still endowed with variable forms
From which all kinds of colours they beget
And vary (by reason that ever it matters much 555
With what seeds, and in what positions joined,
And what the motions that they give and get),
Forthwith most easily thou mayst devise
Why what was black of hue an hour ago

Can of a sudden like the marble gleam,– 560
As ocean, when the high winds have upheaved
Its level plains, is changed to hoary waves
Of marble whiteness: for, thou mayst declare,
That, when the thing we often see as black
Is in its matter then commixed anew, 565
Some atoms rearranged, and some withdrawn,
And added some, 'tis seen forthwith to turn
Glowing and white. But if of azure seeds
Consist the level waters of the deep,
They could in nowise whiten: for however 570
Thou shakest azure seeds, the same can never
Pass into marble hue. But, if the seeds–
Which thus produce the ocean's one pure sheen–
Be now with one hue, now another dyed,
As oft from alien forms and divers shapes 575
A cube's produced all uniform in shape,
'Twould be but natural, even as in the cube
We see the forms to be dissimilar,
That thus we'd see in brightness of the deep
(Or in whatever one pure sheen thou wilt) 580
Colours diverse and all dissimilar.
Besides, the unlike shapes don't thwart the least
The whole in being externally a cube;
But differing hues of things do block and keep
The whole from being of one resultant hue. 585
Then, too, the reason which entices us
At times to attribute colours to the seeds
Falls quite to pieces, since white things are not
Create from white things, nor are black from black,
But evermore they are create from things 590
Of divers colours. Verily, the white
Will rise more readily, is sooner born
Out of no colour, than of black or aught
Which stands in hostile opposition thus.

595

Besides, since colours cannot be, sans light,
And the primordials come not forth to light,
'Tis thine to know they are not clothed with colour–
Truly, what kind of colour could there be
In the viewless dark? Nay, in the light itself 600

A colour changes, gleaming variedly,
When smote by vertical or slanting ray.
Thus in the sunlight shows the down of doves
That circles, garlanding, the nape and throat:
Now it is ruddy with a bright gold-bronze, 605
Now, by a strange sensation it becomes
Green-emerald blended with the coral-red.
The peacock's tail, filled with the copious light,
Changes its colours likewise, when it turns.
Wherefore, since by some blow of light begot, 610
Without such blow these colours can't become.

And since the pupil of the eye receives
Within itself one kind of blow, when said
To feel a white hue, then another kind, 615
When feeling a black or any other hue,
And since it matters nothing with what hue
The things thou touchest be perchance endowed,
But rather with what sort of shape equipped,
'Tis thine to know the atoms need not colour, 620
But render forth sensations, as of touch,
That vary with their varied forms.

Besides,
Since special shapes have not a special colour, 625
And all formations of the primal germs
Can be of any sheen thou wilt, why, then,
Are not those objects which are of them made
Suffused, each kind with colours of every kind?
For then 'twere meet that ravens, as they fly, 630
Should dartle from white pinions a white sheen,
Or swans turn black from seed of black, or be
Of any single varied dye thou wilt.

Again, the more an object's rent to bits, 635
The more thou see its colour fade away
Little by little till 'tis quite extinct;
As happens when the gaudy linen's picked
Shred after shred away: the purple there,
Phoenician red, most brilliant of all dyes, 640
Is lost asunder, ravelled thread by thread;

Hence canst perceive the fragments die away
From out their colour, long ere they depart
Back to the old primordials of things.
And, last, since thou concedest not all bodies 645
Send out a voice or smell, it happens thus
That not to all thou givest sounds and smells.
So, too, since we behold not all with eyes,
'Tis thine to know some things there are as much
Orphaned of colour, as others without smell, 650
And reft of sound; and those the mind alert
No less can apprehend than it can mark
The things that lack some other qualities.

But think not haply that the primal bodies 655
Remain despoiled alone of colour: so,
Are they from warmth dissevered and from cold
And from hot exhalations; and they move,
Both sterile of sound and dry of juice; and throw
Not any odour from their proper bodies. 660
Just as, when undertaking to prepare
A liquid balm of myrrh and marjoram,
And flower of nard, which to our nostrils breathes
Odour of nectar, first of all behooves
Thou seek, as far as find thou may and can, 665
The inodorous olive-oil (which never sends
One whiff of scent to nostrils), that it may
The least debauch and ruin with sharp tang
The odorous essence with its body mixed
And in it seethed. And on the same account 670
The primal germs of things must not be thought
To furnish colour in begetting things,
Nor sound, since pow'rless they to send forth aught
From out themselves, nor any flavour, too,
Nor cold, nor exhalation hot or warm. 675

The rest; yet since these things are mortal all–
The pliant mortal, with a body soft;
The brittle mortal, with a crumbling frame;
The hollow with a porous-all must be 680
Disjoined from the primal elements,
If still we wish under the world to lay

Immortal ground-works, whereupon may rest
The sum of weal and safety, lest for thee
All things return to nothing utterly. 685

Now, too: whate'er we see possessing sense
Must yet confessedly be stablished all
From elements insensate. And those signs,
So clear to all and witnessed out of hand, 690
Do not refute this dictum nor oppose;
But rather themselves do lead us by the hand,
Compelling belief that living things are born
Of elements insensate, as I say.
Sooth, we may see from out the stinking dung 695
Live worms spring up, when, after soaking rains,
The drenched earth rots; and all things change the same:
Lo, change the rivers, the fronds, the gladsome pastures
Into the cattle, the cattle their nature change
Into our bodies, and from our body, oft 700
Grow strong the powers and bodies of wild beasts
And mighty-winged birds. Thus nature changes
All foods to living frames, and procreates
From them the senses of live creatures all,
In manner about as she uncoils in flames 705
Dry logs of wood and turns them all to fire.
And seest not, therefore, how it matters much
After what order are set the primal germs,
And with what other germs they all are mixed,
And what the motions that they give and get? 710

But now, what is't that strikes thy sceptic mind,
Constraining thee to sundry arguments
Against belief that from insensate germs
The sensible is gendered?–Verily, 715
'Tis this: that liquids, earth, and wood, though mixed,
Are yet unable to gender vital sense.
And, therefore, 'twill be well in these affairs
This to remember: that I have not said
Senses are born, under conditions all, 720
From all things absolutely which create
Objects that feel; but much it matters here
Firstly, how small the seeds which thus compose

The feeling thing, then, with what shapes endowed,
And lastly what they in positions be, 725
In motions, in arrangements. Of which facts
Naught we perceive in logs of wood and clods;
And yet even these, when sodden by the rains,
Give birth to wormy grubs, because the bodies
Of matter, from their old arrangements stirred 730
By the new factor, then combine anew
In such a way as genders living things.

Next, they who deem that feeling objects can
From feeling objects be create, and these, 735
In turn, from others that are wont to feel

When soft they make them; for all sense is linked
With flesh, and thews, and veins–and such, we see,
Are fashioned soft and of a mortal frame. 740
Yet be't that these can last forever on:
They'll have the sense that's proper to a part,
Or else be judged to have a sense the same
As that within live creatures as a whole.
But of themselves those parts can never feel, 745
For all the sense in every member back
To something else refers–a severed hand,
Or any other member of our frame,
Itself alone cannot support sensation.
It thus remains they must resemble, then, 750
Live creatures as a whole, to have the power
Of feeling sensation concordant in each part
With the vital sense; and so they're bound to feel
The things we feel exactly as do we.
If such the case, how, then, can they be named 755
The primal germs of things, and how avoid
The highways of destruction?–since they be
Mere living things and living things be all
One and the same with mortal. Grant they could,
Yet by their meetings and their unions all, 760
Naught would result, indeed, besides a throng
And hurly-burly all of living things–
Precisely as men, and cattle, and wild beasts,
By mere conglomeration each with each

Can still beget not anything of new. 765
But if by chance they lose, inside a body,
Their own sense and another sense take on,
What, then, avails it to assign them that
Which is withdrawn thereafter? And besides,
To touch on proof that we pronounced before, 770
Just as we see the eggs of feathered fowls
To change to living chicks, and swarming worms
To bubble forth when from the soaking rains
The earth is sodden, sure, sensations all
Can out of non-sensations be begot. 775

But if one say that sense can so far rise
From non-sense by mutation, or because
Brought forth as by a certain sort of birth,
'Twill serve to render plain to him and prove 780
There is no birth, unless there be before
Some formed union of the elements,
Nor any change, unless they be unite.

In first place, senses can't in body be 785
Before its living nature's been begot,–
Since all its stuff, in faith, is held dispersed
About through rivers, air, and earth, and all
That is from earth created, nor has met
In combination, and, in proper mode, 790
Conjoined into those vital motions which
Kindle the all-perceiving senses–they
That keep and guard each living thing soever.

Again, a blow beyond its nature's strength 795
Shatters forthwith each living thing soe'er,
And on it goes confounding all the sense
Of body and mind. For of the primal germs
Are loosed their old arrangements, and, throughout,
The vital motions blocked,–until the stuff, 800
Shaken profoundly through the frame entire,
Undoes the vital knots of soul from body
And throws that soul, to outward wide-dispersed,
Through all the pores. For what may we surmise
A blow inflicted can achieve besides 805

Shaking asunder and loosening all apart?
It happens also, when less sharp the blow,
The vital motions which are left are wont
Oft to win out–win out, and stop and still
The uncouth tumults gendered by the blow, 810
And call each part to its own courses back,
And shake away the motion of death which now
Begins its own dominion in the body,
And kindle anew the senses almost gone.
For by what other means could they the more 815
Collect their powers of thought and turn again
From very doorways of destruction
Back unto life, rather than pass whereto
They be already well-nigh sped and so
Pass quite away? 820

Again, since pain is there
Where bodies of matter, by some force stirred up,
Through vitals and through joints, within their seats
Quiver and quake inside, but soft delight, 825
When they remove unto their place again:
'Tis thine to know the primal germs can be
Assaulted by no pain, nor from themselves
Take no delight; because indeed they are
Not made of any bodies of first things, 830
Under whose strange new motions they might ache
Or pluck the fruit of any dear new sweet.
And so they must be furnished with no sense.

Once more, if thus, that every living thing 835
May have sensation, needful 'tis to assign
Sense also to its elements, what then
Of those fixed elements from which mankind
Hath been, by their peculiar virtue, formed?
Of verity, they'll laugh aloud, like men, 840
Shaken asunder by a spasm of mirth,
Or sprinkle with dewy tear-drops cheeks and chins,
And have the cunning hardihood to say
Much on the composition of the world,
And in their turn inquire what elements 845
They have themselves,–since, thus the same in kind

As a whole mortal creature, even they
Must also be from other elements,
And then those others from others evermore–
So that thou darest nowhere make a stop. 850
Oho, I'll follow thee until thou grant
The seed (which here thou say'st speaks, laughs, and thinks)
Is yet derived out of other seeds
Which in their turn are doing just the same. 855
But if we see what raving nonsense this,
And that a man may laugh, though not, forsooth,
Compounded out of laughing elements,
And think and utter reason with learn'd speech,
Though not himself compounded, for a fact, 860
Of sapient seeds and eloquent, why, then,
Cannot those things which we perceive to have
Their own sensation be composed as well
Of intermixed seeds quite void of sense?

Infinite Worlds

Once more, we all from seed celestial spring,
To all is that same father, from whom earth,
The fostering mother, as she takes the drops
Of liquid moisture, pregnant bears her broods–
The shining grains, and gladsome shrubs and trees, 5
And bears the human race and of the wild
The generations all, the while she yields
The foods wherewith all feed their frames and lead
The genial life and propagate their kind;
Wherefore she owneth that maternal name, 10
By old desert. What was before from earth,
The same in earth sinks back, and what was sent
From shores of ether, that, returning home,
The vaults of sky receive. Nor thus doth death
So far annihilate things that she destroys 15
The bodies of matter; but she dissipates
Their combinations, and conjoins anew
One element with others; and contrives
That all things vary forms and change their colours
And get sensations and straight give them o'er. 20
And thus may'st know it matters with what others
And in what structure the primordial germs
Are held together, and what motions they
Among themselves do give and get; nor think
That aught we see hither and thither afloat 25
Upon the crest of things, and now a birth
And straightway now a ruin, inheres at rest
Deep in the eternal atoms of the world.

Why, even in these our very verses here 30
It matters much with what and in what order
Each element is set: the same denote
Sky, and the ocean, lands, and streams, and sun;
The same, the grains, and trees, and living things.
And if not all alike, at least the most– 35
But what distinctions by positions wrought!
And thus no less in things themselves, when once
Around are changed the intervals between,
The paths of matter, its connections, weights,
Blows, clashings, motions, order, structure, shapes, 40
The things themselves must likewise changed be.

Now to true reason give thy mind for us.
Since here strange truth is putting forth its might
To hit thee in thine ears, a new aspect 45
Of things to show its front. Yet naught there is
So easy that it standeth not at first
More hard to credit than it after is;
And naught soe'er that's great to such degree,
Nor wonderful so far, but all mankind 50
Little by little abandon their surprise.
Look upward yonder at the bright clear sky
And what it holds–the stars that wander o'er,
The moon, the radiance of the splendour-sun:
Yet all, if now they first for mortals were, 55
If unforeseen now first asudden shown,
What might there be more wonderful to tell,
What that the nations would before have dared
Less to believe might be?–I fancy, naught–
So strange had been the marvel of that sight. 60
The which o'erwearied to behold, to-day
None deigns look upward to those lucent realms.
Then, spew not reason from thy mind away,
Beside thyself because the matter's new,
But rather with keen judgment nicely weigh; 65
And if to thee it then appeareth true,
Render thy hands, or, if 'tis false at last,
Gird thee to combat. For my mind-of-man
Now seeks the nature of the vast Beyond
There on the other side, that boundless sum 70

Which lies without the ramparts of the world,
Toward which the spirit longs to peer afar,
Toward which indeed the swift elan of thought
Flies unencumbered forth.

75

Firstly, we find,
Off to all regions round, on either side,
Above, beneath, throughout the universe
End is there none–as I have taught, as too
The very thing of itself declares aloud, 80
And as from nature of the unbottomed deep
Shines clearly forth. Nor can we once suppose
In any way 'tis likely, (seeing that space
To all sides stretches infinite and free,
And seeds, innumerable in number, in sum 85
Bottomless, there in many a manner fly,
Bestirred in everlasting motion there),
That only this one earth and sky of ours
Hath been create and that those bodies of stuff,
So many, perform no work outside the same; 90
Seeing, moreover, this world too hath been
By nature fashioned, even as seeds of things
By innate motion chanced to clash and cling–
After they'd been in many a manner driven
Together at random, without design, in vain– 95
And as at last those seeds together dwelt,
Which, when together of a sudden thrown,
Should alway furnish the commencements fit
Of mighty things–the earth, the sea, the sky,
And race of living creatures. Thus, I say, 100
Again, again, 'tmust be confessed there are
Such congregations of matter otherwhere,
Like this our world which vasty ether holds
In huge embrace.

105

Besides, when matter abundant
Is ready there, when space on hand, nor object
Nor any cause retards, no marvel 'tis
That things are carried on and made complete,
Perforce. And now, if store of seeds there is 110
So great that not whole life-times of the living

Can count the tale...
And if their force and nature abide the same,
Able to throw the seeds of things together
Into their places, even as here are thrown 115
The seeds together in this world of ours,
'Tmust be confessed in other realms there are
Still other worlds, still other breeds of men,
And other generations of the wild.

120

Hence too it happens in the sum there is
No one thing single of its kind in birth,
And single and sole in growth, but rather it is
One member of some generated race,
Among full many others of like kind. 125
First, cast thy mind abroad upon the living:
Thou'lt find the race of mountain-ranging wild
Even thus to be, and thus the scions of men
To be begot, and lastly the mute flocks
Of scaled fish, and winged frames of birds. 130
Wherefore confess we must on grounds the same
That earth, sun, moon, and ocean, and all else,
Exist not sole and single–rather in number
Exceeding number. Since that deeply set
Old boundary stone of life remains for them 135
No less, and theirs a body of mortal birth
No less, than every kind which here on earth
Is so abundant in its members found.

Which well perceived if thou hold in mind, 140
Then Nature, delivered from every haughty lord,
And forthwith free, is seen to do all things
Herself and through herself of own accord,
Rid of all gods. For–by their holy hearts
Which pass in long tranquillity of peace 145
Untroubled ages and a serene life!–
Who hath the power (I ask), who hath the power
To rule the sum of the immeasurable,
To hold with steady hand the giant reins
Of the unfathomed deep? Who hath the power 150
At once to roll a multitude of skies,
At once to heat with fires ethereal all

The fruitful lands of multitudes of worlds,
To be at all times in all places near,
To stablish darkness by his clouds, to shake 155
The serene spaces of the sky with sound,
And hurl his lightnings,–ha, and whelm how oft
In ruins his own temples, and to rave,
Retiring to the wildernesses, there
At practice with that thunderbolt of his, 160
Which yet how often shoots the guilty by,
And slays the honourable blameless ones!

Ere since the birth-time of the world, ere since
The risen first-born day of sea, earth, sun, 165
Have many germs been added from outside,
Have many seeds been added round about,
Which the great All, the while it flung them on,
Brought hither, that from them the sea and lands
Could grow more big, and that the house of heaven 170
Might get more room and raise its lofty roofs
Far over earth, and air arise around.
For bodies all, from out all regions, are
Divided by blows, each to its proper thing,
And all retire to their own proper kinds: 175
The moist to moist retires; earth gets increase
From earthy body; and fires, as on a forge,
Beat out new fire; and ether forges ether;
Till nature, author and ender of the world,
Hath led all things to extreme bound of growth: 180
As haps when that which hath been poured inside
The vital veins of life is now no more
Than that which ebbs within them and runs off.
This is the point where life for each thing ends;
This is the point where nature with her powers 185
Curbs all increase. For whatsoe'er thou seest
Grow big with glad increase, and step by step
Climb upward to ripe age, these to themselves
Take in more bodies than they send from selves,
Whilst still the food is easily infused 190
Through all the veins, and whilst the things are not
So far expanded that they cast away
Such numerous atoms as to cause a waste

Greater than nutriment whereby they wax.
For 'tmust be granted, truly, that from things 195
Many a body ebbeth and runs off;
But yet still more must come, until the things
Have touched development's top pinnacle;
Then old age breaks their powers and ripe strength
And falls away into a worser part. 200
For ever the ampler and more wide a thing,
As soon as ever its augmentation ends,
It scatters abroad forthwith to all sides round
More bodies, sending them from out itself.
Nor easily now is food disseminate 205
Through all its veins; nor is that food enough
To equal with a new supply on hand
Those plenteous exhalations it gives off.
Thus, fairly, all things perish, when with ebbing
They're made less dense and when from blows without 210
They are laid low; since food at last will fail
Extremest eld, and bodies from outside
Cease not with thumping to undo a thing
And overmaster by infesting blows.

215

Thus, too, the ramparts of the mighty world
On all sides round shall taken be by storm,
And tumble to wrack and shivered fragments down.
For food it is must keep things whole, renewing;
'Tis food must prop and give support to all,– 220
But to no purpose, since nor veins suffice
To hold enough, nor nature ministers
As much as needful. And even now 'tis thus:
Its age is broken and the earth, outworn
With many parturitions, scarce creates 225
The little lives–she who created erst
All generations and gave forth at birth
Enormous bodies of wild beasts of old.
For never, I fancy, did a golden cord
From off the firmament above let down 230
The mortal generations to the fields;
Nor sea, nor breakers pounding on the rocks
Created them; but earth it was who bore–
The same to-day who feeds them from herself.

Besides, herself of own accord, she first 235
The shining grains and vineyards of all joy
Created for mortality; herself
Gave the sweet fruitage and the pastures glad,
Which now to-day yet scarcely wax in size,
Even when aided by our toiling arms. 240
We break the ox, and wear away the strength
Of sturdy farm-hands; iron tools to-day
Barely avail for tilling of the fields,
So niggardly they grudge our harvestings,
So much increase our labour. Now to-day 245
The aged ploughman, shaking of his head,
Sighs o'er and o'er that labours of his hands
Have fallen out in vain, and, as he thinks
How present times are not as times of old,
Often he praises the fortunes of his sire, 250
And crackles, prating, how the ancient race,
Fulfilled with piety, supported life
With simple comfort in a narrow plot,
Since, man for man, the measure of each field
Was smaller far i' the old days. And, again, 255
The gloomy planter of the withered vine
Rails at the season's change and wearies heaven,
Nor grasps that all of things by sure degrees
Are wasting away and going to the tomb,
Outworn by venerable length of life. 260

Book Three

Proem

O thou who first uplifted in such dark
So clear a torch aloft, who first shed light
Upon the profitable ends of man,
O thee I follow, glory of the Greeks,
And set my footsteps squarely planted now 5
Even in the impress and the marks of thine–
Less like one eager to dispute the palm,
More as one craving out of very love
That I may copy thee!–for how should swallow
Contend with swans or what compare could be 10
In a race between young kids with tumbling legs
And the strong might of the horse? Our father thou,
And finder-out of truth, and thou to us
Suppliest a father's precepts; and from out
Those scriven leaves of thine, renowned soul 15
(Like bees that sip of all in flowery wolds),
We feed upon thy golden sayings all–
Golden, and ever worthiest endless life.
For soon as ever thy planning thought that sprang
From god-like mind begins its loud proclaim 20
Of nature's courses, terrors of the brain
Asunder flee, the ramparts of the world
Dispart away, and through the void entire
I see the movements of the universe.
Rises to vision the majesty of gods, 25
And their abodes of everlasting calm

Which neither wind may shake nor rain-cloud splash,
Nor snow, congealed by sharp frosts, may harm
With its white downfall: ever, unclouded sky
O'er roofs, and laughs with far-diffused light. 30
And nature gives to them their all, nor aught
May ever pluck their peace of mind away.
But nowhere to my vision rise no more
The vaults of Acheron, though the broad earth
Bars me no more from gazing down o'er all 35
Which under our feet is going on below
Along the void. O, here in these affairs
Some new divine delight and trembling awe
Takes hold through me, that thus by power of thine
Nature, so plain and manifest at last, 40
Hath been on every side laid bare to man!

And since I've taught already of what sort
The seeds of all things are, and how, distinct
In divers forms, they flit of own accord, 45
Stirred with a motion everlasting on,
And in what mode things be from them create,
Now, after such matters, should my verse, meseems,
Make clear the nature of the mind and soul,
And drive that dread of Acheron without, 50
Headlong, which so confounds our human life
Unto its deeps, pouring o'er all that is
The black of death, nor leaves not anything
To prosper–a liquid and unsullied joy.
For as to what men sometimes will affirm: 55
That more than Tartarus (the realm of death)
They fear diseases and a life of shame,
And know the substance of the soul is blood,
Or rather wind (if haply thus their whim),
And so need naught of this our science, then 60
Thou well may'st note from what's to follow now
That more for glory do they braggart forth
Than for belief. For mark these very same:
Exiles from country, fugitives afar
From sight of men, with charges foul attaint, 65
Abased with every wretchedness, they yet
Live, and where'er the wretches come, they yet

Make the ancestral sacrifices there,
Butcher the black sheep, and to gods below
Offer the honours, and in bitter case 70
Turn much more keenly to religion.
Wherefore, it's surer testing of a man
In doubtful perils–mark him as he is
Amid adversities; for then alone
Are the true voices conjured from his breast, 75
The mask off-stripped, reality behind.
And greed, again, and the blind lust of honours
Which force poor wretches past the bounds of law,
And, oft allies and ministers of crime,
To push through nights and days with hugest toil 80
To rise untrammelled to the peaks of power–
These wounds of life in no mean part are kept
Festering and open by this fright of death.
For ever we see fierce Want and foul Disgrace
Dislodged afar from secure life and sweet, 85
Like huddling Shapes before the doors of death.
And whilst, from these, men wish to scape afar,
Driven by false terror, and afar remove,
With civic blood a fortune they amass,
They double their riches, greedy, heapers-up 90
Of corpse on corpse they have a cruel laugh
For the sad burial of a brother-born,
And hatred and fear of tables of their kin.
Likewise, through this same terror, envy oft
Makes them to peak because before their eyes 95
That man is lordly, that man gazed upon
Who walks begirt with honour glorious,
Whilst they in filth and darkness roll around;
Some perish away for statues and a name,
And oft to that degree, from fright of death, 100
Will hate of living and beholding light
Take hold on humankind that they inflict
Their own destruction with a gloomy heart–
Forgetful that this fear is font of cares,
This fear the plague upon their sense of shame, 105
And this that breaks the ties of comradry
And oversets all reverence and faith,
Mid direst slaughter. For long ere to-day

Often were traitors to country and dear parents
Through quest to shun the realms of Acheron. 110
For just as children tremble and fear all
In the viewless dark, so even we at times
Dread in the light so many things that be
No whit more fearsome than what children feign,
Shuddering, will be upon them in the dark. 115
This terror, then, this darkness of the mind,
Not sunrise with its flaring spokes of light,
Nor glittering arrows of morning sun disperse,
But only nature's aspect and her law.

Nature and Composition of the Mind

First, then, I say, the mind which oft we call
The intellect, wherein is seated life's
Counsel and regimen, is part no less
Of man than hand and foot and eyes are parts
Of one whole breathing creature. [But some hold] 5
That sense of mind is in no fixed part seated,
But is of body some one vital state,–
Named "harmony" by Greeks, because thereby
We live with sense, though intellect be not
In any part: as oft the body is said 10
To have good health (when health, however, 's not
One part of him who has it), so they place
The sense of mind in no fixed part of man.
Mightily, diversly, meseems they err.
Often the body palpable and seen 15
Sickens, while yet in some invisible part
We feel a pleasure; oft the other way,
A miserable in mind feels pleasure still
Throughout his body–quite the same as when
A foot may pain without a pain in head. 20
Besides, when these our limbs are given o'er
To gentle sleep and lies the burdened frame
At random void of sense, a something else
Is yet within us, which upon that time
Bestirs itself in many a wise, receiving 25
All motions of joy and phantom cares of heart.
Now, for to see that in man's members dwells
Also the soul, and body ne'er is wont
To feel sensation by a "harmony"

Take this in chief: the fact that life remains 30
Oft in our limbs, when much of body's gone;
Yet that same life, when particles of heat,
Though few, have scattered been, and through the mouth
Air has been given forth abroad, forthwith
Forever deserts the veins, and leaves the bones. 35
Thus mayst thou know that not all particles
Perform like parts, nor in like manner all
Are props of weal and safety: rather those–
The seeds of wind and exhalations warm–
Take care that in our members life remains. 40
Therefore a vital heat and wind there is
Within the very body, which at death
Deserts our frames. And so, since nature of mind
And even of soul is found to be, as 'twere,
A part of man, give over "harmony"– 45
Name to musicians brought from Helicon,–
Unless themselves they filched it otherwise,
To serve for what was lacking name till then.
Whate'er it be, they're welcome to it–thou,
Hearken my other maxims. 50

Mind and soul,
I say, are held conjoined one with other,
And form one single nature of themselves;
But chief and regnant through the frame entire 55
Is still that counsel which we call the mind,
And that cleaves seated in the midmost breast.
Here leap dismay and terror; round these haunts
Be blandishments of joys; and therefore here
The intellect, the mind. The rest of soul, 60
Throughout the body scattered, but obeys–
Moved by the nod and motion of the mind.
This, for itself, sole through itself, hath thought;
This for itself hath mirth, even when the thing
That moves it, moves nor soul nor body at all. 65
And as, when head or eye in us is smit
By assailing pain, we are not tortured then
Through all the body, so the mind alone
Is sometimes smitten, or livens with a joy,
Whilst yet the soul's remainder through the limbs 70

And through the frame is stirred by nothing new.
But when the mind is moved by shock more fierce,
We mark the whole soul suffering all at once
Along man's members: sweats and pallors spread
Over the body, and the tongue is broken, 75
And fails the voice away, and ring the ears,
Mists blind the eyeballs, and the joints collapse,–
Aye, men drop dead from terror of the mind.
Hence, whoso will can readily remark
That soul conjoined is with mind, and, when 80
'Tis strook by influence of the mind, forthwith
In turn it hits and drives the body too.

And this same argument establisheth
That nature of mind and soul corporeal is: 85
For when 'tis seen to drive the members on,
To snatch from sleep the body, and to change
The countenance, and the whole state of man
To rule and turn,–what yet could never be
Sans contact, and sans body contact fails– 90
Must we not grant that mind and soul consist
Of a corporeal nature?–And besides
Thou markst that likewise with this body of ours
Suffers the mind and with our body feels.
If the dire speed of spear that cleaves the bones 95
And bares the inner thews hits not the life,
Yet follows a fainting and a foul collapse,
And, on the ground, dazed tumult in the mind,
And whiles a wavering will to rise afoot.
So nature of mind must be corporeal, since 100
From stroke and spear corporeal 'tis in throes.

Now, of what body, what components formed
Is this same mind I will go on to tell.
First, I aver, 'tis superfine, composed 105
Of tiniest particles–that such the fact
Thou canst perceive, if thou attend, from this:
Nothing is seen to happen with such speed
As what the mind proposes and begins;
Therefore the same bestirs itself more swiftly 110
Than aught whose nature's palpable to eyes.

But what's so agile must of seeds consist
Most round, most tiny, that they may be moved,
When hit by impulse slight. So water moves,
In waves along, at impulse just the least– 115
Being create of little shapes that roll;
But, contrariwise, the quality of honey
More stable is, its liquids more inert,
More tardy its flow; for all its stock of matter
Cleaves more together, since, indeed, 'tis made 120
Of atoms not so smooth, so fine, and round.
For the light breeze that hovers yet can blow
High heaps of poppy-seed away for thee
Downward from off the top; but, contrariwise,
A pile of stones or spiny ears of wheat 125
It can't at all. Thus, in so far as bodies
Are small and smooth, is their mobility;
But, contrariwise, the heavier and more rough,
The more immovable they prove. Now, then,
Since nature of mind is movable so much, 130
Consist it must of seeds exceeding small
And smooth and round. Which fact once known to thee,
Good friend, will serve thee opportune in else.
This also shows the nature of the same,
How nice its texture, in how small a space 135
'Twould go, if once compacted as a pellet:
When death's unvexed repose gets hold on man
And mind and soul retire, thou markest there
From the whole body nothing ta'en in form,
Nothing in weight. Death grants ye everything, 140
But vital sense and exhalation hot.
Thus soul entire must be of smallmost seeds,
Twined through the veins, the vitals, and the thews,
Seeing that, when 'tis from whole body gone,
The outward figuration of the limbs 145
Is unimpaired and weight fails not a whit.
Just so, when vanished the bouquet of wine,
Or when an unguent's perfume delicate
Into the winds away departs, or when
From any body savour's gone, yet still 150
The thing itself seems minished naught to eyes,
Thereby, nor aught abstracted from its weight–

No marvel, because seeds many and minute
Produce the savours and the redolence
In the whole body of the things. And so, 155
Again, again, nature of mind and soul
'Tis thine to know created is of seeds
The tiniest ever, since at flying-forth
It beareth nothing of the weight away.
160

Yet fancy not its nature simple so.
For an impalpable aura, mixed with heat,
Deserts the dying, and heat draws off the air;
And heat there's none, unless commixed with air:
For, since the nature of all heat is rare, 165
Athrough it many seeds of air must move.
Thus nature of mind is triple; yet those all
Suffice not for creating sense–since mind
Accepteth not that aught of these can cause
Sense-bearing motions, and much less the thoughts 170
A man revolves in mind. So unto these
Must added be a somewhat, and a fourth;
That somewhat's altogether void of name;
Than which existeth naught more mobile, naught
More an impalpable, of elements 175
More small and smooth and round. That first transmits
Sense-bearing motions through the frame, for that
Is roused the first, composed of little shapes;
Thence heat and viewless force of wind take up
The motions, and thence air, and thence all things 180
Are put in motion; the blood is strook, and then
The vitals all begin to feel, and last
To bones and marrow the sensation comes–
Pleasure or torment. Nor will pain for naught
Enter so far, nor a sharp ill seep through, 185
But all things be perturbed to that degree
That room for life will fail, and parts of soul
Will scatter through the body's every pore.
Yet as a rule, almost upon the skin
These motion all are stopped, and this is why 190
We have the power to retain our life.

Now in my eagerness to tell thee how

They are commixed, through what unions fit
They function so, my country's pauper-speech 195
Constrains me sadly. As I can, however,
I'll touch some points and pass. In such a wise
Course these primordials 'mongst one another
With inter-motions that no one can be
From other sundered, nor its agency 200
Perform, if once divided by a space;
Like many powers in one body they work.
As in the flesh of any creature still
Is odour and savour and a certain warmth,
And yet from all of these one bulk of body 205
Is made complete, so, viewless force of wind
And warmth and air, commingled, do create
One nature, by that mobile energy
Assisted which from out itself to them
Imparts initial motion, whereby first 210
Sense-bearing motion along the vitals springs.
For lurks this essence far and deep and under,
Nor in our body is aught more shut from view,
And 'tis the very soul of all the soul.
And as within our members and whole frame 215
The energy of mind and power of soul
Is mixed and latent, since create it is
Of bodies small and few, so lurks this fourth,
This essence void of name, composed of small,
And seems the very soul of all the soul, 220
And holds dominion o'er the body all.
And by like reason wind and air and heat
Must function so, commingled through the frame,
And now the one subside and now another
In interchange of dominance, that thus 225
From all of them one nature be produced,
Lest heat and wind apart, and air apart,
Make sense to perish, by disseverment.
There is indeed in mind that heat it gets
When seething in rage, and flashes from the eyes 230
More swiftly fire; there is, again, that wind,
Much, and so cold, companion of all dread,
Which rouses the shudder in the shaken frame;
There is no less that state of air composed,

Making the tranquil breast, the serene face. 235
But more of hot have they whose restive hearts,
Whose minds of passion quickly seethe in rage–
Of which kind chief are fierce abounding lions,
Who often with roaring burst the breast o'erwrought,
Unable to hold the surging wrath within; 240
But the cold mind of stags has more of wind,
And speedier through their inwards rouses up
The icy currents which make their members quake.
But more the oxen live by tranquil air,
Nor e'er doth smoky torch of wrath applied, 245
O'erspreading with shadows of a darkling murk,
Rouse them too far; nor will they stiffen stark,
Pierced through by icy javelins of fear;
But have their place half-way between the two–
Stags and fierce lions. Thus the race of men: 250
Though training make them equally refined,
It leaves those pristine vestiges behind
Of each mind's nature. Nor may we suppose
Evil can e'er be rooted up so far
That one man's not more given to fits of wrath, 255
Another's not more quickly touched by fear,
A third not more long-suffering than he should.
And needs must differ in many things besides
The varied natures and resulting habits
Of humankind–of which not now can I 260
Expound the hidden causes, nor find names
Enough for all the divers shapes of those
Primordials whence this variation springs.
But this meseems I'm able to declare:
Those vestiges of natures left behind 265
Which reason cannot quite expel from us
Are still so slight that naught prevents a man
From living a life even worthy of the gods.

So then this soul is kept by all the body, 270
Itself the body's guard, and source of weal:
For they with common roots cleave each to each,
Nor can be torn asunder without death.
Not easy 'tis from lumps of frankincense
To tear their fragrance forth, without its nature 275

Perishing likewise: so, not easy 'tis
From all the body nature of mind and soul
To draw away, without the whole dissolved.
With seeds so intertwined even from birth,
They're dowered conjointly with a partner-life; 280
No energy of body or mind, apart,
Each of itself without the other's power,
Can have sensation; but our sense, enkindled
Along the vitals, to flame is blown by both
With mutual motions. Besides the body alone 285
Is nor begot nor grows, nor after death
Seen to endure. For not as water at times
Gives off the alien heat, nor is thereby
Itself destroyed, but unimpaired remains–
Not thus, I say, can the deserted frame 290
Bear the dissevering of its joined soul,
But, rent and ruined, moulders all away.
Thus the joint contact of the body and soul
Learns from their earliest age the vital motions,
Even when still buried in the mother's womb; 295
So no dissevering can hap to them,
Without their bane and ill. And thence mayst see
That, as conjoined is their source of weal,
Conjoined also must their nature be.

300

If one, moreover, denies that body feel,
And holds that soul, through all the body mixed,
Takes on this motion which we title "sense,"
He battles in vain indubitable facts:
For who'll explain what body's feeling is, 305
Except by what the public fact itself
Has given and taught us?"But when soul is parted,
Body's without all sense." True!–loses what
Was even in its life-time not its own;
And much beside it loses, when soul's driven 310
Forth from that life-time. Or, to say that eyes
Themselves can see no thing, but through the same
The mind looks forth, as out of opened doors,
Is–a hard saying; since the feel in eyes
Says the reverse. For this itself draws on 315
And forces into the pupils of our eyes

Our consciousness. And note the case when often
We lack the power to see refulgent things,
Because our eyes are hampered by their light–
With a mere doorway this would happen not; 320
For, since it is our very selves that see,
No open portals undertake the toil.
Besides, if eyes of ours but act as doors,
Methinks that, were our sight removed, the mind
Ought then still better to behold a thing– 325
When even the door-posts have been cleared away.

Herein in these affairs nowise take up
What honoured sage, Democritus, lays down–
That proposition, that primordials 330
Of body and mind, each super-posed on each,
Vary alternately and interweave
The fabric of our members. For not only
Are the soul-elements smaller far than those
Which this our body and inward parts compose, 335
But also are they in their number less,
And scattered sparsely through our frame. And thus
This canst thou guarantee: soul's primal germs
Maintain between them intervals as large
At least as are the smallest bodies, which, 340
When thrown against us, in our body rouse
Sense-bearing motions. Hence it comes that we
Sometimes don't feel alighting on our frames
The clinging dust, or chalk that settles soft;
Nor mists of night, nor spider's gossamer 345
We feel against us, when, upon our road,
Its net entangles us, nor on our head
The dropping of its withered garmentings;
Nor bird-feathers, nor vegetable down,
Flying about, so light they barely fall; 350
Nor feel the steps of every crawling thing,
Nor each of all those footprints on our skin
Of midges and the like. To that degree
Must many primal germs be stirred in us
Ere once the seeds of soul that through our frame 355
Are intermingled 'gin to feel that those
Primordials of the body have been strook,

And ere, in pounding with such gaps between,
They clash, combine and leap apart in turn.

360

But mind is more the keeper of the gates,
Hath more dominion over life than soul.
For without intellect and mind there's not
One part of soul can rest within our frame
Least part of time; companioning, it goes 365
With mind into the winds away, and leaves
The icy members in the cold of death.
But he whose mind and intellect abide
Himself abides in life. However much
The trunk be mangled, with the limbs lopped off, 370
The soul withdrawn and taken from the limbs,
Still lives the trunk and draws the vital air.
Even when deprived of all but all the soul,
Yet will it linger on and cleave to life,–
Just as the power of vision still is strong, 375
If but the pupil shall abide unharmed,
Even when the eye around it's sorely rent–
Provided only thou destroyest not
Wholly the ball, but, cutting round the pupil,
Leavest that pupil by itself behind– 380
For more would ruin sight. But if that centre,
That tiny part of eye, be eaten through,
Forthwith the vision fails and darkness comes,
Though in all else the unblemished ball be clear.
'Tis by like compact that the soul and mind 385
Are each to other bound forevermore.

The Soul is Mortal

Now come: that thou mayst able be to know
That minds and the light souls of all that live
Have mortal birth and death, I will go on
Verses to build meet for thy rule of life,
Sought after long, discovered with sweet toil. 5
But under one name I'd have thee yoke them both;
And when, for instance, I shall speak of soul,
Teaching the same to be but mortal, think
Thereby I'm speaking also of the mind–
Since both are one, a substance inter-joined. 10
First, then, since I have taught how soul exists
A subtle fabric, of particles minute,
Made up from atoms smaller much than those
Of water's liquid damp, or fog, or smoke,
So in mobility it far excels, 15
More prone to move, though strook by lighter cause
Even moved by images of smoke or fog–
As where we view, when in our sleeps we're lulled,
The altars exhaling steam and smoke aloft–
For, beyond doubt, these apparitions come 20
To us from outward. Now, then, since thou seest,
Their liquids depart, their waters flow away,
When jars are shivered, and since fog and smoke
Depart into the winds away, believe
The soul no less is shed abroad and dies 25
More quickly far, more quickly is dissolved
Back to its primal bodies, when withdrawn
From out man's members it has gone away.
For, sure, if body (container of the same

Like as a jar), when shivered from some cause, 30
And rarefied by loss of blood from veins,
Cannot for longer hold the soul, how then
Thinkst thou it can be held by any air–
A stuff much rarer than our bodies be?

35

Besides we feel that mind to being comes
Along with body, with body grows and ages.
For just as children totter round about
With frames infirm and tender, so there follows
A weakling wisdom in their minds; and then, 40
Where years have ripened into robust powers,
Counsel is also greater, more increased
The power of mind; thereafter, where already
The body's shattered by master-powers of eld,
And fallen the frame with its enfeebled powers, 45
Thought hobbles, tongue wanders, and the mind gives way;
All fails, all's lacking at the selfsame time.
Therefore it suits that even the soul's dissolved,
Like smoke, into the lofty winds of air;
Since we behold the same to being come 50
Along with body and grow, and, as I've taught,
Crumble and crack, therewith outworn by eld.

Then, too, we see, that, just as body takes
Monstrous diseases and the dreadful pain, 55
So mind its bitter cares, the grief, the fear;
Wherefore it tallies that the mind no less
Partaker is of death; for pain and disease
Are both artificers of death,–as well
We've learned by the passing of many a man ere now. 60
Nay, too, in diseases of body, often the mind
Wanders afield; for 'tis beside itself,
And crazed it speaks, or many a time it sinks,
With eyelids closing and a drooping nod,
In heavy drowse, on to eternal sleep; 65
From whence nor hears it any voices more,
Nor able is to know the faces here
Of those about him standing with wet cheeks
Who vainly call him back to light and life.
Wherefore mind too, confess we must, dissolves, 70

Seeing, indeed, contagions of disease
Enter into the same. Again, O why,
When the strong wine has entered into man,
And its diffused fire gone round the veins,
Why follows then a heaviness of limbs, 75
A tangle of the legs as round he reels,
A stuttering tongue, an intellect besoaked,
Eyes all aswim, and hiccups, shouts, and brawls,
And whatso else is of that ilk?–Why this?–
If not that violent and impetuous wine 80
Is wont to confound the soul within the body?
But whatso can confounded be and balked,
Gives proof, that if a hardier cause got in,
'Twould hap that it would perish then, bereaved
Of any life thereafter. And, moreover, 85
Often will some one in a sudden fit,
As if by stroke of lightning, tumble down
Before our eyes, and sputter foam, and grunt,
Blither, and twist about with sinews taut,
Gasp up in starts, and weary out his limbs 90
With tossing round. No marvel, since distract
Through frame by violence of disease.

Confounds, he foams, as if to vomit soul,
As on the salt sea boil the billows round 95
Under the master might of winds. And now
A groan's forced out, because his limbs are griped,
But, in the main, because the seeds of voice
Are driven forth and carried in a mass
Outwards by mouth, where they are wont to go, 100
And have a builded highway. He becomes
Mere fool, since energy of mind and soul
Confounded is, and, as I've shown, to-riven,
Asunder thrown, and torn to pieces all
By the same venom. But, again, where cause 105
Of that disease has faced about, and back
Retreats sharp poison of corrupted frame
Into its shadowy lairs, the man at first
Arises reeling, and gradually comes back
To all his senses and recovers soul. 110
Thus, since within the body itself of man

The mind and soul are by such great diseases
Shaken, so miserably in labour distraught,
Why, then, believe that in the open air,
Without a body, they can pass their life, 115
Immortal, battling with the master winds?
And, since we mark the mind itself is cured,
Like the sick body, and restored can be
By medicine, this is forewarning too
That mortal lives the mind. For proper it is 120
That whosoe'er begins and undertakes
To alter the mind, or meditates to change
Any another nature soever, should add
New parts, or readjust the order given,
Or from the sum remove at least a bit. 125
But what's immortal willeth for itself
Its parts be nor increased, nor rearranged,
Nor any bit soever flow away:
For change of anything from out its bounds
Means instant death of that which was before. 130
Ergo, the mind, whether in sickness fallen,
Or by the medicine restored, gives signs,
As I have taught, of its mortality.
So surely will a fact of truth make head
'Gainst errors' theories all, and so shut off 135
All refuge from the adversary, and rout
Error by two-edged confutation.

And since the mind is of a man one part,
Which in one fixed place remains, like ears, 140
And eyes, and every sense which pilots life;
And just as hand, or eye, or nose, apart,
Severed from us, can neither feel nor be,
But in the least of time is left to rot,
Thus mind alone can never be, without 145
The body and the man himself, which seems,
As 'twere the vessel of the same–or aught
Whate'er thou'lt feign as yet more closely joined:
Since body cleaves to mind by surest bonds.

150

Again, the body's and the mind's live powers
Only in union prosper and enjoy;

For neither can nature of mind, alone of self
Sans body, give the vital motions forth;
Nor, then, can body, wanting soul, endure 155
And use the senses. Verily, as the eye,
Alone, up-rended from its roots, apart
From all the body, can peer about at naught,
So soul and mind it seems are nothing able,
When by themselves. No marvel, because, commixed 160
Through veins and inwards, and through bones and thews,
Their elements primordial are confined
By all the body, and own no power free
To bound around through interspaces big,
Thus, shut within these confines, they take on 165
Motions of sense, which, after death, thrown out
Beyond the body to the winds of air,
Take on they cannot–and on this account,
Because no more in such a way confined.
For air will be a body, be alive, 170
If in that air the soul can keep itself,
And in that air enclose those motions all
Which in the thews and in the body itself
A while ago 'twas making. So for this,
Again, again, I say confess we must, 175
That, when the body's wrappings are unwound,
And when the vital breath is forced without,
The soul, the senses of the mind dissolve,–
Since for the twain the cause and ground of life
Is in the fact of their conjoined estate. 180

Once more, since body's unable to sustain
Division from the soul, without decay
And obscene stench, how canst thou doubt but that
The soul, uprisen from the body's deeps, 185
Has filtered away, wide-drifted like a smoke,
Or that the changed body crumbling fell
With ruin so entire, because, indeed,
Its deep foundations have been moved from place,
The soul out-filtering even through the frame, 190
And through the body's every winding way
And orifice? And so by many means
Thou'rt free to learn that nature of the soul

Hath passed in fragments out along the frame,
And that 'twas shivered in the very body 195
Ere ever it slipped abroad and swam away
Into the winds of air. For never a man
Dying appears to feel the soul go forth
As one sure whole from all his body at once,
Nor first come up the throat and into mouth; 200
But feels it failing in a certain spot,
Even as he knows the senses too dissolve
Each in its own location in the frame.
But were this mind of ours immortal mind,
Dying 'twould scarce bewail a dissolution, 205
But rather the going, the leaving of its coat,
Like to a snake. Wherefore, when once the body
Hath passed away, admit we must that soul,
Shivered in all that body, perished too.
Nay, even when moving in the bounds of life, 210
Often the soul, now tottering from some cause,
Craves to go out, and from the frame entire
Loosened to be; the countenance becomes
Flaccid, as if the supreme hour were there;
And flabbily collapse the members all 215
Against the bloodless trunk–the kind of case
We see when we remark in common phrase,
"That man's quite gone," or "fainted dead away";
And where there's now a bustle of alarm,
And all are eager to get some hold upon 220
The man's last link of life. For then the mind
And all the power of soul are shook so sore,
And these so totter along with all the frame,
That any cause a little stronger might
Dissolve them altogether.–Why, then, doubt 225
That soul, when once without the body thrust,
There in the open, an enfeebled thing,
Its wrappings stripped away, cannot endure
Not only through no everlasting age,
But even, indeed, through not the least of time? 230

Then, too, why never is the intellect,
The counselling mind, begotten in the head,
The feet, the hands, instead of cleaving still

To one sole seat, to one fixed haunt, the breast, 235
If not that fixed places be assigned
For each thing's birth, where each, when 'tis create,
Is able to endure, and that our frames
Have such complex adjustments that no shift
In order of our members may appear? 240
To that degree effect succeeds to cause,
Nor is the flame once wont to be create
In flowing streams, nor cold begot in fire.

Besides, if nature of soul immortal be, 245
And able to feel, when from our frame disjoined,
The same, I fancy, must be thought to be
Endowed with senses five,–nor is there way
But this whereby to image to ourselves
How under-souls may roam in Acheron. 250
Thus painters and the elder race of bards
Have pictured souls with senses so endowed.
But neither eyes, nor nose, nor hand, alone
Apart from body can exist for soul,
Nor tongue nor ears apart. And hence indeed 255
Alone by self they can nor feel nor be.

And since we mark the vital sense to be
In the whole body, all one living thing,
If of a sudden a force with rapid stroke 260
Should slice it down the middle and cleave in twain,
Beyond a doubt likewise the soul itself,
Divided, dissevered, asunder will be flung
Along with body. But what severed is
And into sundry parts divides, indeed 265
Admits it owns no everlasting nature.
We hear how chariots of war, areek
With hurly slaughter, lop with flashing scythes
The limbs away so suddenly that there,
Fallen from the trunk, they quiver on the earth, 270
The while the mind and powers of the man
Can feel no pain, for swiftness of his hurt,
And sheer abandon in the zest of battle:
With the remainder of his frame he seeks
Anew the battle and the slaughter, nor marks 275

How the swift wheels and scythes of ravin have dragged
Off with the horses his left arm and shield;
Nor other how his right has dropped away,
Mounting again and on. A third attempts
With leg dismembered to arise and stand, 280
Whilst, on the ground hard by, the dying foot
Twitches its spreading toes. And even the head,
When from the warm and living trunk lopped off,
Keeps on the ground the vital countenance
And open eyes, until 't has rendered up 285
All remnants of the soul. Nay, once again:
If, when a serpent's darting forth its tongue,
And lashing its tail, thou gettest chance to hew
With axe its length of trunk to many parts,
Thou'lt see each severed fragment writhing round 290
With its fresh wound, and spattering up the sod,
And there the fore-part seeking with the jaws
After the hinder, with bite to stop the pain.
So shall we say that these be souls entire
In all those fractions?–but from that 'twould follow 295
One creature'd have in body many souls.
Therefore, the soul, which was indeed but one,
Has been divided with the body too:
Each is but mortal, since alike is each
Hewn into many parts. Again, how often 300
We view our fellow going by degrees,
And losing limb by limb the vital sense;
First nails and fingers of the feet turn blue,
Next die the feet and legs, then o'er the rest
Slow crawl the certain footsteps of cold death. 305
And since this nature of the soul is torn,
Nor mounts away, as at one time, entire,
We needs must hold it mortal. But perchance
If thou supposest that the soul itself
Can inward draw along the frame, and bring 310
Its parts together to one place, and so
From all the members draw the sense away,
Why, then, that place in which such stock of soul
Collected is, should greater seem in sense.
But since such place is nowhere, for a fact, 315
As said before, 'tis rent and scattered forth,

And so goes under. Or again, if now
I please to grant the false, and say that soul
Can thus be lumped within the frames of those
Who leave the sunshine, dying bit by bit, 320
Still must the soul as mortal be confessed;
Nor aught it matters whether to wrack it go,
Dispersed in the winds, or, gathered in a mass
From all its parts, sink down to brutish death,
Since more and more in every region sense 325
Fails the whole man, and less and less of life
In every region lingers.

And besides,
If soul immortal is, and winds its way 330
Into the body at the birth of man,
Why can we not remember something, then,
Of life-time spent before? why keep we not
Some footprints of the things we did of, old?
But if so changed hath been the power of mind, 335
That every recollection of things done
Is fallen away, at no o'erlong remove
Is that, I trow, from what we mean by death.
Wherefore 'tis sure that what hath been before
Hath died, and what now is is now create. 340

Moreover, if after the body hath been built
Our mind's live powers are wont to be put in,
Just at the moment that we come to birth,
And cross the sills of life, 'twould scarcely fit 345
For them to live as if they seemed to grow
Along with limbs and frame, even in the blood,
But rather as in a cavern all alone.
(Yet all the body duly throngs with sense.)
But public fact declares against all this: 350
For soul is so entwined through the veins,
The flesh, the thews, the bones, that even the teeth
Share in sensation, as proven by dull ache,
By twinge from icy water, or grating crunch
Upon a stone that got in mouth with bread. 355
Wherefore, again, again, souls must be thought
Nor void of birth, nor free from law of death;

Nor, if, from outward, in they wound their way,
Could they be thought as able so to cleave
To these our frames, nor, since so interwove, 360
Appears it that they're able to go forth
Unhurt and whole and loose themselves unscathed
From all the thews, articulations, bones.
But, if perchance thou thinkest that the soul,
From outward winding in its way, is wont 365
To seep and soak along these members ours,
Then all the more 'twill perish, being thus
With body fused–for what will seep and soak
Will be dissolved and will therefore die.
For just as food, dispersed through all the pores 370
Of body, and passed through limbs and all the frame,
Perishes, supplying from itself the stuff
For other nature, thus the soul and mind,
Though whole and new into a body going,
Are yet, by seeping in, dissolved away, 375
Whilst, as through pores, to all the frame there pass
Those particles from which created is
This nature of mind, now ruler of our body,
Born from that soul which perished, when divided
Along the frame. Wherefore it seems that soul 380
Hath both a natal and funeral hour.

Besides are seeds of soul there left behind
In the breathless body, or not? If there they are,
It cannot justly be immortal deemed, 385
Since, shorn of some parts lost, 'thas gone away:
But if, borne off with members uncorrupt,
'Thas fled so absolutely all away
It leaves not one remainder of itself
Behind in body, whence do cadavers, then, 390
From out their putrid flesh exhale the worms,
And whence does such a mass of living things,
Boneless and bloodless, o'er the bloated frame
Bubble and swarm? But if perchance thou thinkest
That souls from outward into worms can wind, 395
And each into a separate body come,
And reckonest not why many thousand souls
Collect where only one has gone away,

Here is a point, in sooth, that seems to need
Inquiry and a putting to the test: 400
Whether the souls go on a hunt for seeds
Of worms wherewith to build their dwelling places,
Or enter bodies ready-made, as 'twere.
But why themselves they thus should do and toil
'Tis hard to say, since, being free of body, 405
They flit around, harassed by no disease,
Nor cold nor famine; for the body labours
By more of kinship to these flaws of life,
And mind by contact with that body suffers
So many ills. But grant it be for them 410
However useful to construct a body
To which to enter in, 'tis plain they can't.
Then, souls for self no frames nor bodies make,
Nor is there how they once might enter in
To bodies ready-made–for they cannot 415
Be nicely interwoven with the same,
And there'll be formed no interplay of sense
Common to each.

Again, why is't there goes 420
Impetuous rage with lion's breed morose,
And cunning with foxes, and to deer why given
The ancestral fear and tendency to flee,
And why in short do all the rest of traits
Engender from the very start of life 425
In the members and mentality, if not
Because one certain power of mind that came
From its own seed and breed waxes the same
Along with all the body? But were mind
Immortal, were it wont to change its bodies, 430
How topsy-turvy would earth's creatures act!
The Hyrcan hound would flee the onset oft
Of antlered stag, the scurrying hawk would quake
Along the winds of air at the coming dove,
And men would dote, and savage beasts be wise; 435
For false the reasoning of those that say
Immortal mind is changed by change of body–
For what is changed dissolves, and therefore dies.
For parts are re-disposed and leave their order;

Wherefore they must be also capable 440
Of dissolution through the frame at last,
That they along with body perish all.
But should some say that always souls of men
Go into human bodies, I will ask:
How can a wise become a dullard soul? 445
And why is never a child's a prudent soul?
And the mare's filly why not trained so well
As sturdy strength of steed? We may be sure
They'll take their refuge in the thought that mind
Becomes a weakling in a weakling frame. 450
Yet be this so, 'tis needful to confess
The soul but mortal, since, so altered now
Throughout the frame, it loses the life and sense
It had before. Or how can mind wax strong
Coequally with body and attain 455
The craved flower of life, unless it be
The body's colleague in its origins?
Or what's the purport of its going forth
From aged limbs?–fears it, perhaps, to stay,
Pent in a crumbled body? Or lest its house, 460
Outworn by venerable length of days,
May topple down upon it? But indeed
For an immortal perils are there none.

Again, at parturitions of the wild 465
And at the rites of Love, that souls should stand
Ready hard by seems ludicrous enough–
Immortals waiting for their mortal limbs
In numbers innumerable, contending madly
Which shall be first and chief to enter in!– 470
Unless perchance among the souls there be
Such treaties stablished that the first to come
Flying along, shall enter in the first,
And that they make no rivalries of strength!

475

Again, in ether can't exist a tree,
Nor clouds in ocean deeps, nor in the fields
Can fishes live, nor blood in timber be,
Nor sap in boulders: fixed and arranged
Where everything may grow and have its place. 480

Thus nature of mind cannot arise alone
Without the body, nor exist afar
From thews and blood. But if 'twere possible,
Much rather might this very power of mind
Be in the head, the shoulders or the heels, 485
And, born in any part soever, yet
In the same man, in the same vessel abide.
But since within this body even of ours
Stands fixed and appears arranged sure
Where soul and mind can each exist and grow, 490
Deny we must the more that they can have
Duration and birth, wholly outside the frame.
For, verily, the mortal to conjoin
With the eternal, and to feign they feel
Together, and can function each with each, 495
Is but to dote: for what can be conceived
Of more unlike, discrepant, ill-assorted,
Than something mortal in a union joined
With an immortal and a secular
To bear the outrageous tempests? 500

Then, again,
Whatever abides eternal must indeed
Either repel all strokes, because 'tis made
Of solid body, and permit no entrance 505
Of aught with power to sunder from within
The parts compact–as are those seeds of stuff
Whose nature we've exhibited before;
Or else be able to endure through time
For this: because they are from blows exempt, 510
As is the void, the which abides untouched,
Unsmit by any stroke; or else because
There is no room around, whereto things can,
As 'twere, depart in dissolution all,–
Even as the sum of sums eternal is, 515
Without or place beyond whereto things may
Asunder fly, or bodies which can smite,
And thus dissolve them by the blows of might.

But if perchance the soul's to be adjudged 520
Immortal, mainly on ground 'tis kept secure

In vital forces–either because there come
Never at all things hostile to its weal,
Or else because what come somehow retire,
Repelled or ere we feel the harm they work, 525

For, lo, besides that, when the frame's diseased,
Soul sickens too, there cometh, many a time,
That which torments it with the things to be,
Keeps it in dread, and wearies it with cares; 530
And even when evil acts are of the past,
Still gnaw the old transgressions bitterly.
Add, too, that frenzy, peculiar to the mind,
And that oblivion of the things that were;
Add its submergence in the murky waves 535
Of drowse and torpor.

Folly of the Fear of Death

Therefore death to us
Is nothing, nor concerns us in the least,
Since nature of mind is mortal evermore.
And just as in the ages gone before
We felt no touch of ill, when all sides round 5
To battle came the Carthaginian host,
And the times, shaken by tumultuous war,
Under the aery coasts of arching heaven
Shuddered and trembled, and all humankind
Doubted to which the empery should fall 10
By land and sea, thus when we are no more,
When comes that sundering of our body and soul
Through which we're fashioned to a single state,
Verily naught to us, us then no more,
Can come to pass, naught move our senses then– 15
No, not if earth confounded were with sea,
And sea with heaven. But if indeed do feel
The nature of mind and energy of soul,
After their severance from this body of ours,
Yet nothing 'tis to us who in the bonds 20
And wedlock of the soul and body live,
Through which we're fashioned to a single state.
And, even if time collected after death
The matter of our frames and set it all
Again in place as now, and if again 25
To us the light of life were given, O yet
That process too would not concern us aught,
When once the self-succession of our sense
Has been asunder broken. And now and here,

Little enough we're busied with the selves 30
We were aforetime, nor, concerning them,
Suffer a sore distress. For shouldst thou gaze
Backwards across all yesterdays of time
The immeasurable, thinking how manifold
The motions of matter are, then couldst thou well 35
Credit this too: often these very seeds
(From which we are to-day) of old were set
In the same order as they are to-day–
Yet this we can't to consciousness recall
Through the remembering mind. For there hath been 40
An interposed pause of life, and wide
Have all the motions wandered everywhere
From these our senses. For if woe and ail
Perchance are toward, then the man to whom
The bane can happen must himself be there 45
At that same time. But death precludeth this,
Forbidding life to him on whom might crowd
Such irk and care; and granted 'tis to know:
Nothing for us there is to dread in death,
No wretchedness for him who is no more, 50
The same estate as if ne'er born before,
When death immortal hath ta'en the mortal life.

Hence, where thou seest a man to grieve because
When dead he rots with body laid away, 55
Or perishes in flames or jaws of beasts,
Know well: he rings not true, and that beneath
Still works an unseen sting upon his heart,
However he deny that he believes.
His shall be aught of feeling after death. 60
For he, I fancy, grants not what he says,
Nor what that presupposes, and he fails
To pluck himself with all his roots from life
And cast that self away, quite unawares
Feigning that some remainder's left behind. 65
For when in life one pictures to oneself
His body dead by beasts and vultures torn,
He pities his state, dividing not himself
Therefrom, removing not the self enough
From the body flung away, imagining 70

Himself that body, and projecting there
His own sense, as he stands beside it: hence
He grieves that he is mortal born, nor marks
That in true death there is no second self
Alive and able to sorrow for self destroyed, 75
Or stand lamenting that the self lies there
Mangled or burning. For if it an evil is
Dead to be jerked about by jaw and fang
Of the wild brutes, I see not why 'twere not
Bitter to lie on fires and roast in flames, 80
Or suffocate in honey, and, reclined
On the smooth oblong of an icy slab,
Grow stiff in cold, or sink with load of earth
Down-crushing from above.

85

"Thee now no more
The joyful house and best of wives shall welcome,
Nor little sons run up to snatch their kisses
And touch with silent happiness thy heart.
Thou shalt not speed in undertakings more, 90
Nor be the warder of thine own no more.
Poor wretch," they say, "one hostile hour hath ta'en
Wretchedly from thee all life's many guerdons,"
But add not, "yet no longer unto thee
Remains a remnant of desire for them" 95
If this they only well perceived with mind
And followed up with maxims, they would free
Their state of man from anguish and from fear.
"O even as here thou art, aslumber in death,
So shalt thou slumber down the rest of time, 100
Released from every harrying pang. But we,
We have bewept thee with insatiate woe,
Standing beside whilst on the awful pyre
Thou wert made ashes; and no day shall take
For us the eternal sorrow from the breast." 105
But ask the mourner what's the bitterness
That man should waste in an eternal grief,
If, after all, the thing's but sleep and rest?
For when the soul and frame together are sunk
In slumber, no one then demands his self 110
Or being. Well, this sleep may be forever,

Without desire of any selfhood more,
For all it matters unto us asleep.
Yet not at all do those primordial germs
Roam round our members, at that time, afar 115
From their own motions that produce our senses–
Since, when he's startled from his sleep, a man
Collects his senses. Death is, then, to us
Much less–if there can be a less than that
Which is itself a nothing: for there comes 120
Hard upon death a scattering more great
Of the throng of matter, and no man wakes up
On whom once falls the icy pause of life.

This too, O often from the soul men say, 125
Along their couches holding of the cups,
With faces shaded by fresh wreaths awry:
"Brief is this fruit of joy to paltry man,
Soon, soon departed, and thereafter, no,
It may not be recalled."–As if, forsooth, 130
It were their prime of evils in great death
To parch, poor tongues, with thirst and arid drought,
Or chafe for any lack.

Once more, if Nature 135
Should of a sudden send a voice abroad,
And her own self inveigh against us so:
"Mortal, what hast thou of such grave concern
That thou indulgest in too sickly plaints?
Why this bemoaning and beweeping death? 140
For if thy life aforetime and behind
To thee was grateful, and not all thy good
Was heaped as in sieve to flow away
And perish unavailingly, why not,
Even like a banqueter, depart the halls, 145
Laden with life? why not with mind content
Take now, thou fool, thy unafflicted rest?
But if whatever thou enjoyed hath been
Lavished and lost, and life is now offence,
Why seekest more to add–which in its turn 150
Will perish foully and fall out in vain?
O why not rather make an end of life,

Of labour? For all I may devise or find
To pleasure thee is nothing: all things are
The same forever. Though not yet thy body 155
Wrinkles with years, nor yet the frame exhausts
Outworn, still things abide the same, even if
Thou goest on to conquer all of time
With length of days, yea, if thou never diest"–
What were our answer, but that Nature here 160
Urges just suit and in her words lays down
True cause of action? Yet should one complain,
Riper in years and elder, and lament,
Poor devil, his death more sorely than is fit,
Then would she not, with greater right, on him 165
Cry out, inveighing with a voice more shrill:
"Off with thy tears, and choke thy whines, buffoon!
Thou wrinklest–after thou hast had the sum
Of the guerdons of life; yet, since thou cravest ever
What's not at hand, contemning present good, 170
That life has slipped away, unperfected
And unavailing unto thee. And now,
Or ere thou guessed it, death beside thy head
Stands–and before thou canst be going home
Sated and laden with the goodly feast. 175
But now yield all that's alien to thine age,–
Up, with good grace! make room for sons: thou must."
Justly, I fancy, would she reason thus,
Justly inveigh and gird: since ever the old
Outcrowded by the new gives way, and ever 180
The one thing from the others is repaired.
Nor no man is consigned to the abyss
Of Tartarus, the black. For stuff must be,
That thus the after-generations grow,–
Though these, their life completed, follow thee; 185
And thus like thee are generations all–
Already fallen, or some time to fall.
So one thing from another rises ever;
And in fee-simple life is given to none,
But unto all mere usufruct. 190

Look back:
Nothing to us was all fore-passed eld

Of time the eternal, ere we had a birth.
And Nature holds this like a mirror up 195
Of time-to-be when we are dead and gone.
And what is there so horrible appears?
Now what is there so sad about it all?
Is't not serener far than any sleep?
200

And, verily, those tortures said to be
In Acheron, the deep, they all are ours
Here in this life. No Tantalus, benumbed
With baseless terror, as the fables tell,
Fears the huge boulder hanging in the air: 205
But, rather, in life an empty dread of Gods
Urges mortality, and each one fears
Such fall of fortune as may chance to him.
Nor eat the vultures into Tityus
Prostrate in Acheron, nor can they find, 210
Forsooth, throughout eternal ages, aught
To pry around for in that mighty breast.
However hugely he extend his bulk–
Who hath for outspread limbs not acres nine,
But the whole earth–he shall not able be 215
To bear eternal pain nor furnish food
From his own frame forever. But for us
A Tityus is he whom vultures rend
Prostrate in love, whom anxious anguish eats,
Whom troubles of any unappeased desires 220
Asunder rip. We have before our eyes
Here in this life also a Sisyphus
In him who seeketh of the populace
The rods, the axes fell, and evermore
Retires a beaten and a gloomy man. 225
For to seek after power–an empty name,
Nor given at all–and ever in the search
To endure a world of toil, O this it is
To shove with shoulder up the hill a stone
Which yet comes rolling back from off the top, 230
And headlong makes for levels of the plain.
Then to be always feeding an ingrate mind,
Filling with good things, satisfying never–
As do the seasons of the year for us,

When they return and bring their progenies 235
And varied charms, and we are never filled
With the fruits of life–O this, I fancy, 'tis
To pour, like those young virgins in the tale,
Waters into a sieve, unfilled forever.

240

Cerberus and Furies, and that Lack of Light

Tartarus, out-belching from his mouth the surge
Of horrible heat–the which are nowhere, nor
Indeed can be: but in this life is fear 245
Of retributions just and expiations
For evil acts: the dungeon and the leap
From that dread rock of infamy, the stripes,
The executioners, the oaken rack,
The iron plates, bitumen, and the torch. 250
And even though these are absent, yet the mind,
With a fore-fearing conscience, plies its goads
And burns beneath the lash, nor sees meanwhile
What terminus of ills, what end of pine
Can ever be, and feareth lest the same 255
But grow more heavy after death. Of truth,
The life of fools is Acheron on earth.

This also to thy very self sometimes
Repeat thou mayst: "Lo, even good Ancus left 260
The sunshine with his eyes, in divers things
A better man than thou, O worthless hind;
And many other kings and lords of rule
Thereafter have gone under, once who swayed
O'er mighty peoples. And he also, he– 265
Who whilom paved a highway down the sea,
And gave his legionaries thoroughfare
Along the deep, and taught them how to cross
The pools of brine afoot, and did contemn,
Trampling upon it with his cavalry, 270
The bellowings of ocean–poured his soul
From dying body, as his light was ta'en.
And Scipio's son, the thunderbolt of war,
Horror of Carthage, gave his bones to earth,
Like to the lowliest villein in the house. 275

Add finders-out of sciences and arts;
Add comrades of the Heliconian dames,
Among whom Homer, sceptered o'er them all,
Now lies in slumber sunken with the rest.
Then, too, Democritus, when ripened eld 280
Admonished him his memory waned away,
Of own accord offered his head to death.
Even Epicurus went, his light of life
Run out, the man in genius who o'er-topped
The human race, extinguishing all others, 285
As sun, in ether arisen, all the stars.
Wilt thou, then, dally, thou complain to go?–
For whom already life's as good as dead,
Whilst yet thou livest and lookest?–who in sleep
Wastest thy life–time's major part, and snorest 290
Even when awake, and ceasest not to see
The stuff of dreams, and bearest a mind beset
By baseless terror, nor discoverest oft
What's wrong with thee, when, like a sotted wretch,
Thou'rt jostled along by many crowding cares, 295
And wanderest reeling round, with mind aswim."

If men, in that same way as on the mind
They feel the load that wearies with its weight,
Could also know the causes whence it comes, 300
And why so great the heap of ill on heart,
O not in this sort would they live their life,
As now so much we see them, knowing not
What 'tis they want, and seeking ever and ever
A change of place, as if to drop the burden. 305
The man who sickens of his home goes out,
Forth from his splendid halls, and straight–returns,
Feeling i'faith no better off abroad.
He races, driving his Gallic ponies along,
Down to his villa, madly,–as in haste 310
To hurry help to a house afire.–At once
He yawns, as soon as foot has touched the threshold,
Or drowsily goes off in sleep and seeks
Forgetfulness, or maybe bustles about
And makes for town again. In such a way 315
Each human flees himself–a self in sooth,

As happens, he by no means can escape;
And willy-nilly he cleaves to it and loathes,
Sick, sick, and guessing not the cause of ail.
Yet should he see but that, O chiefly then, 320
Leaving all else, he'd study to divine
The nature of things, since here is in debate
Eternal time and not the single hour,
Mortal's estate in whatsoever remains
After great death. 325

And too, when all is said,
What evil lust of life is this so great
Subdues us to live, so dreadfully distraught
In perils and alarms? one fixed end 330
Of life abideth for mortality;
Death's not to shun, and we must go to meet.
Besides we're busied with the same devices,
Ever and ever, and we are at them ever,
And there's no new delight that may be forged 335
By living on. But whilst the thing we long for
Is lacking, that seems good above all else;
Thereafter, when we've touched it, something else
We long for; ever one equal thirst of life
Grips us agape. And doubtful 'tis what fortune 340
The future times may carry, or what be
That chance may bring, or what the issue next
Awaiting us. Nor by prolonging life
Take we the least away from death's own time,
Nor can we pluck one moment off, whereby 345
To minish the aeons of our state of death.
Therefore, O man, by living on, fulfil
As many generations as thou may:
Eternal death shall there be waiting still;
And he who died with light of yesterday 350
Shall be no briefer time in death's No-more
Than he who perished months or years before.

Book Four

Proem

I wander afield, thriving in sturdy thought,
Through unpathed haunts of the Pierides,
Trodden by step of none before. I joy
To come on undefiled fountains there,
To drain them deep; I joy to pluck new flowers, 5
To seek for this my head a signal crown
From regions where the Muses never yet
Have garlanded the temples of a man:
First, since I teach concerning mighty things,
And go right on to loose from round the mind 10
The tightened coils of dread religion;
Next, since, concerning themes so dark, I frame
Song so pellucid, touching all throughout
Even with the Muses' charm–which, as 'twould seem,
Is not without a reasonable ground: 15
For as physicians, when they seek to give
Young boys the nauseous wormwood, first do touch
The brim around the cup with the sweet juice
And yellow of the honey, in order that
The thoughtless age of boyhood be cajoled 20
As far as the lips, and meanwhile swallow down
The wormwood's bitter draught, and, though befooled,
Be yet not merely duped, but rather thus
Grow strong again with recreated health:
So now I too (since this my doctrine seems 25
In general somewhat woeful unto those

Who've had it not in hand, and since the crowd
Starts back from it in horror) have desired
To expound our doctrine unto thee in song
Soft-speaking and Pierian, and, as 'twere, 30
To touch it with sweet honey of the Muse–
If by such method haply I might hold
The mind of thee upon these lines of ours,
Till thou dost learn the nature of all things
And understandest their utility. 35

Existence and Character of the Images

But since I've taught already of what sort
The seeds of all things are, and how distinct
In divers forms they flit of own accord,
Stirred with a motion everlasting on,
And in what mode things be from them create, 5
And since I've taught what the mind's nature is,
And of what things 'tis with the body knit
And thrives in strength, and by what mode uptorn
That mind returns to its primordials,
Now will I undertake an argument– 10
One for these matters of supreme concern–
That there exist those somewhats which we call
The images of things: these, like to films
Scaled off the utmost outside of the things,
Flit hither and thither through the atmosphere, 15
And the same terrify our intellects,
Coming upon us waking or in sleep,
When oft we peer at wonderful strange shapes
And images of people lorn of light,
Which oft have horribly roused us when we lay 20
In slumber–that haply nevermore may we
Suppose that souls get loose from Acheron,
Or shades go floating in among the living,
Or aught of us is left behind at death,
When body and mind, destroyed together, each 25
Back to its own primordials goes away.

And thus I say that effigies of things,
And tenuous shapes from off the things are sent,
From off the utmost outside of the things, 30
Which are like films or may be named a rind,
Because the image bears like look and form
With whatso body has shed it fluttering forth–
A fact thou mayst, however dull thy wits,
Well learn from this: mainly, because we see 35
Even 'mongst visible objects many be
That send forth bodies, loosely some diffused–
Like smoke from oaken logs and heat from fires–
And some more interwoven and condensed–
As when the locusts in the summertime 40
Put off their glossy tunics, or when calves
At birth drop membranes from their body's surface,
Or when, again, the slippery serpent doffs
Its vestments 'mongst the thorns–for oft we see
The breres augmented with their flying spoils: 45
Since such takes place, 'tis likewise certain too
That tenuous images from things are sent,
From off the utmost outside of the things.
For why those kinds should drop and part from things,
Rather than others tenuous and thin, 50
No power has man to open mouth to tell;
Especially, since on outsides of things
Are bodies many and minute which could,
In the same order which they had before,
And with the figure of their form preserved, 55
Be thrown abroad, and much more swiftly too,
Being less subject to impediments,
As few in number and placed along the front.
For truly many things we see discharge
Their stuff at large, not only from their cores 60
Deep-set within, as we have said above,
But from their surfaces at times no less–
Their very colours too. And commonly
The awnings, saffron, red and dusky blue,
Stretched overhead in mighty theatres, 65
Upon their poles and cross-beams fluttering,
Have such an action quite; for there they dye
And make to undulate with their every hue

The circled throng below, and all the stage,
And rich attire in the patrician seats. 70
And ever the more the theatre's dark walls
Around them shut, the more all things within
Laugh in the bright suffusion of strange glints,
The daylight being withdrawn. And therefore, since
The canvas hangings thus discharge their dye 75
From off their surface, things in general must
Likewise their tenuous effigies discharge,
Because in either case they are off-thrown
From off the surface. So there are indeed
Such certain prints and vestiges of forms 80
Which flit around, of subtlest texture made,
Invisible, when separate, each and one.
Again, all odour, smoke, and heat, and such
Streams out of things diffusedly, because,
Whilst coming from the deeps of body forth 85
And rising out, along their bending path
They're torn asunder, nor have gateways straight
Wherethrough to mass themselves and struggle abroad.
But contrariwise, when such a tenuous film
Of outside colour is thrown off, there's naught 90
Can rend it, since 'tis placed along the front
Ready to hand. Lastly those images
Which to our eyes in mirrors do appear,
In water, or in any shining surface,
Must be, since furnished with like look of things, 95
Fashioned from images of things sent out.
There are, then, tenuous effigies of forms,
Like unto them, which no one can divine
When taken singly, which do yet give back,
When by continued and recurrent discharge 100
Expelled, a picture from the mirrors' plane.
Nor otherwise, it seems, can they be kept
So well conserved that thus be given back
Figures so like each object.

105

Now then, learn
How tenuous is the nature of an image.
And in the first place, since primordials be
So far beneath our senses, and much less

E'en than those objects which begin to grow 110
Too small for eyes to note, learn now in few
How nice are the beginnings of all things–
That this, too, I may yet confirm in proof:
First, living creatures are sometimes so small
That even their third part can nowise be seen; 115
Judge, then, the size of any inward organ–
What of their sphered heart, their eyes, their limbs,
The skeleton?–How tiny thus they are!
And what besides of those first particles
Whence soul and mind must fashioned be?–Seest not 120
How nice and how minute? Besides, whatever
Exhales from out its body a sharp smell–
The nauseous absinth, or the panacea,
Strong southernwood, or bitter centaury–
If never so lightly with thy [fingers] twain 125
Perchance [thou touch] a one of them

Then why not rather know that images
Flit hither and thither, many, in many modes,
Bodiless and invisible? 130

But lest
Haply thou holdest that those images
Which come from objects are the sole that flit,
Others indeed there be of own accord 135
Begot, self-formed in earth's aery skies,
Which, moulded to innumerable shapes,
Are borne aloft, and, fluid as they are,
Cease not to change appearance and to turn
Into new outlines of all sorts of forms; 140
As we behold the clouds grow thick on high
And smirch the serene vision of the world,
Stroking the air with motions. For oft are seen
The giants' faces flying far along
And trailing a spread of shadow; and at times 145
The mighty mountains and mountain-sundered rocks
Going before and crossing on the sun,
Whereafter a monstrous beast dragging amain
And leading in the other thunderheads.
Now [hear] how easy and how swift they be 150

Engendered, and perpetually flow off
From things and gliding pass away....

For ever every outside streams away
From off all objects, since discharge they may; 155
And when this outside reaches other things,
As chiefly glass, it passes through; but where
It reaches the rough rocks or stuff of wood,
There 'tis so rent that it cannot give back
An image. But when gleaming objects dense, 160
As chiefly mirrors, have been set before it,
Nothing of this sort happens. For it can't
Go, as through glass, nor yet be rent–its safety,
By virtue of that smoothness, being sure.
'Tis therefore that from them the images 165
Stream back to us; and howso suddenly
Thou place, at any instant, anything
Before a mirror, there an image shows;
Proving that ever from a body's surface
Flow off thin textures and thin shapes of things. 170
Thus many images in little time
Are gendered; so their origin is named
Rightly a speedy. And even as the sun
Must send below, in little time, to earth
So many beams to keep all things so full 175
Of light incessant; thus, on grounds the same,
From things there must be borne, in many modes,
To every quarter round, upon the moment,
The many images of things; because
Unto whatever face of things we turn 180
The mirror, things of form and hue the same
Respond. Besides, though but a moment since
Serenest was the weather of the sky,
So fiercely sudden is it foully thick
That ye might think that round about all murk 185
Had parted forth from Acheron and filled
The mighty vaults of sky–so grievously,
As gathers thus the storm-clouds' gruesome night,
Do faces of black horror hang on high–
Of which how small a part an image is 190
There's none to tell or reckon out in words.

Now come; with what swift motion they are borne,
These images, and what the speed assigned
To them across the breezes swimming on–
So that o'er lengths of space a little hour 195
Alone is wasted, toward whatever region
Each with its divers impulse tends–I'll tell
In verses sweeter than they many are;
Even as the swan's slight note is better far
Than that dispersed clamour of the cranes 200
Among the southwind's aery clouds. And first,
One oft may see that objects which are light
And made of tiny bodies are the swift;
In which class is the sun's light and his heat,
Since made from small primordial elements 205
Which, as it were, are forward knocked along
And through the interspaces of the air
To pass delay not, urged by blows behind;
For light by light is instantly supplied
And gleam by following gleam is spurred and driven. 210
Thus likewise must the images have power
Through unimaginable space to speed
Within a point of time,–first, since a cause
Exceeding small there is, which at their back
Far forward drives them and propels, where, too, 215
They're carried with such winged lightness on;
And, secondly, since furnished, when sent off,
With texture of such rareness that they can
Through objects whatsoever penetrate
And ooze, as 'twere, through intervening air. 220
Besides, if those fine particles of things
Which from so deep within are sent abroad,
As light and heat of sun, are seen to glide
And spread themselves through all the space of heaven
Upon one instant of the day, and fly 225
O'er sea and lands and flood the heaven, what then
Of those which on the outside stand prepared,
When they're hurled off with not a thing to check
Their going out? Dost thou not see indeed
How swifter and how farther must they go 230
And speed through manifold the length of space
In time the same that from the sun the rays

O'erspread the heaven? This also seems to be
Example chief and true with what swift speed
The images of things are borne about: 235
That soon as ever under open skies
Is spread the shining water, all at once,
If stars be out in heaven, upgleam from earth,
Serene and radiant in the water there,
The constellations of the universe– 240
Now seest thou not in what a point of time
An image from the shores of ether falls
Unto the shores of earth? Wherefore, again,
And yet again, 'tis needful to confess
With wondrous... 245

The Senses and Mental Pictures

Bodies that strike the eyes, awaking sight.
From certain things flow odours evermore,
As cold from rivers, heat from sun, and spray
From waves of ocean, eater-out of walls
Around the coasts. Nor ever cease to flit 5
The varied voices, sounds athrough the air.
Then too there comes into the mouth at times
The wet of a salt taste, when by the sea
We roam about; and so, whene'er we watch
The wormword being mixed, its bitter stings. 10
To such degree from all things is each thing
Borne streamingly along, and sent about
To every region round; and nature grants
Nor rest nor respite of the onward flow,
Since 'tis incessantly we feeling have, 15
And all the time are suffered to descry
And smell all things at hand, and hear them sound.
Besides, since shape examined by our hands
Within the dark is known to be the same
As that by eyes perceived within the light 20
And lustrous day, both touch and sight must be
By one like cause aroused. So, if we test
A square and get its stimulus on us
Within the dark, within the light what square
Can fall upon our sight, except a square 25
That images the things? Wherefore it seems
The source of seeing is in images,
Nor without these can anything be viewed.

Now these same films I name are borne about 30
And tossed and scattered into regions all.
But since we do perceive alone through eyes,
It follows hence that whitherso we turn
Our sight, all things do strike against it there
With form and hue. And just how far from us 35
Each thing may be away, the image yields
To us the power to see and chance to tell:
For when 'tis sent, at once it shoves ahead
And drives along the air that's in the space
Betwixt it and our eyes. And thus this air 40
All glides athrough our eyeballs, and, as 'twere,
Brushes athrough our pupils and thuswise
Passes across. Therefore it comes we see
How far from us each thing may be away,
And the more air there be that's driven before, 45
And too the longer be the brushing breeze
Against our eyes, the farther off removed
Each thing is seen to be: forsooth, this work
With mightily swift order all goes on,
So that upon one instant we may see 50
What kind the object and how far away.

Nor over-marvellous must this be deemed
In these affairs that, though the films which strike
Upon the eyes cannot be singly seen, 55
The things themselves may be perceived. For thus
When the wind beats upon us stroke by stroke
And when the sharp cold streams, 'tis not our wont
To feel each private particle of wind
Or of that cold, but rather all at once; 60
And so we see how blows affect our body,
As if one thing were beating on the same
And giving us the feel of its own body
Outside of us. Again, whene'er we thump
With finger-tip upon a stone, we touch 65
But the rock's surface and the outer hue,
Nor feel that hue by contact–rather feel
The very hardness deep within the rock.

Now come, and why beyond a looking-glass 70

An image may be seen, perceive. For seen
It soothly is, removed far within.
'Tis the same sort as objects peered upon
Outside in their true shape, whene'er a door
Yields through itself an open peering-place, 75
And lets us see so many things outside
Beyond the house. Also that sight is made
By a twofold twin air: for first is seen
The air inside the door-posts; next the doors,
The twain to left and right; and afterwards 80
A light beyond comes brushing through our eyes,
Then other air, then objects peered upon
Outside in their true shape. And thus, when first
The image of the glass projects itself,
As to our gaze it comes, it shoves ahead 85
And drives along the air that's in the space
Betwixt it and our eyes, and brings to pass
That we perceive the air ere yet the glass.
But when we've also seen the glass itself,
Forthwith that image which from us is borne 90
Reaches the glass, and there thrown back again
Comes back unto our eyes, and driving rolls
Ahead of itself another air, that then
'Tis this we see before itself, and thus
It looks so far removed behind the glass. 95
Wherefore again, again, there's naught for wonder

In those which render from the mirror's plane
A vision back, since each thing comes to pass
By means of the two airs. Now, in the glass 100
The right part of our members is observed
Upon the left, because, when comes the image
Hitting against the level of the glass,
'Tis not returned unshifted; but forced off
Backwards in line direct and not oblique,– 105
Exactly as whoso his plaster-mask
Should dash, before 'twere dry, on post or beam,
And it should straightway keep, at clinging there,
Its shape, reversed, facing him who threw,
And so remould the features it gives back: 110
It comes that now the right eye is the left,

The left the right. An image too may be
From mirror into mirror handed on,
Until of idol-films even five or six
Have thus been gendered. For whatever things 115
Shall hide back yonder in the house, the same,
However far removed in twisting ways,
May still be all brought forth through bending paths
And by these several mirrors seen to be
Within the house, since nature so compels 120
All things to be borne backward and spring off
At equal angles from all other things.
To such degree the image gleams across
From mirror unto mirror; where 'twas left
It comes to be the right, and then again 125
Returns and changes round unto the left.
Again, those little sides of mirrors curved
Proportionate to the bulge of our own flank
Send back to us their idols with the right
Upon the right; and this is so because 130
Either the image is passed on along
From mirror unto mirror, and thereafter,
When twice dashed off, flies back unto ourselves;
Or else the image wheels itself around,
When once unto the mirror it has come, 135
Since the curved surface teaches it to turn
To usward. Further, thou might'st well believe
That these film-idols step along with us
And set their feet in unison with ours
And imitate our carriage, since from that 140
Part of a mirror whence thou hast withdrawn
Straightway no images can be returned.

Further, our eye-balls tend to flee the bright
And shun to gaze thereon; the sun even blinds, 145
If thou goest on to strain them unto him,
Because his strength is mighty, and the films
Heavily downward from on high are borne
Through the pure ether and the viewless winds,
And strike the eyes, disordering their joints. 150
So piecing lustre often burns the eyes,
Because it holdeth many seeds of fire

Which, working into eyes, engender pain.
Again, whatever jaundiced people view
Becomes wan-yellow, since from out their bodies 155
Flow many seeds wan-yellow forth to meet
The films of things, and many too are mixed
Within their eye, which by contagion paint
All things with sallowness. Again, we view
From dark recesses things that stand in light, 160
Because, when first has entered and possessed
The open eyes this nearer darkling air,
Swiftly the shining air and luminous
Followeth in, which purges then the eyes
And scatters asunder of that other air 165
The sable shadows, for in large degrees
This air is nimbler, nicer, and more strong.
And soon as ever 'thas filled and oped with light
The pathways of the eyeballs, which before
Black air had blocked, there follow straightaway 170
Those films of things out-standing in the light,
Provoking vision–what we cannot do
From out the light with objects in the dark,
Because that denser darkling air behind
Followeth in, and fills each aperture 175
And thus blockades the pathways of the eyes
That there no images of any things
Can be thrown in and agitate the eyes.

And when from far away we do behold 180
The squared towers of a city, oft
Rounded they seem,–on this account because
Each distant angle is perceived obtuse,
Or rather it is not perceived at all;
And perishes its blow nor to our gaze 185
Arrives its stroke, since through such length of air
Are borne along the idols that the air
Makes blunt the idol of the angle's point
By numerous collidings. When thuswise
The angles of the tower each and all 190
Have quite escaped the sense, the stones appear
As rubbed and rounded on a turner's wheel–
Yet not like objects near and truly round,

But with a semblance to them, shadowily.
Likewise, our shadow in the sun appears 195
To move along and follow our own steps
And imitate our carriage–if thou thinkest
Air that is thus bereft of light can walk,
Following the gait and motion of mankind.
For what we use to name a shadow, sure 200
Is naught but air deprived of light. No marvel:
Because the earth from spot to spot is reft
Progressively of light of sun, whenever
In moving round we get within its way,
While any spot of earth by us abandoned 205
Is filled with light again, on this account
It comes to pass that what was body's shadow
Seems still the same to follow after us
In one straight course. Since, evermore pour in
New lights of rays, and perish then the old, 210
Just like the wool that's drawn into the flame.
Therefore the earth is easily spoiled of light
And easily refilled and from herself
Washeth the black shadows quite away.
215

And yet in this we don't at all concede
That eyes be cheated. For their task it is
To note in whatsoever place be light,
In what be shadow: whether or no the gleams
Be still the same, and whether the shadow which 220
Just now was here is that one passing thither,
Or whether the facts be what we said above,
'Tis after all the reasoning of mind
That must decide; nor can our eyeballs know
The nature of reality. And so 225
Attach thou not this fault of mind to eyes,
Nor lightly think our senses everywhere
Are tottering. The ship in which we sail
Is borne along, although it seems to stand;
The ship that bides in roadstead is supposed 230
There to be passing by. And hills and fields
Seem fleeing fast astern, past which we urge
The ship and fly under the bellying sails.
The stars, each one, do seem to pause, affixed

To the ethereal caverns, though they all 235
Forever are in motion, rising out
And thence revisiting their far descents
When they have measured with their bodies bright
The span of heaven. And likewise sun and moon
Seem biding in a roadstead,–objects which, 240
As plain fact proves, are really borne along.
Between two mountains far away aloft
From midst the whirl of waters open lies
A gaping exit for the fleet, and yet
They seem conjoined in a single isle. 245
When boys themselves have stopped their spinning round,
The halls still seem to whirl and posts to reel,
Until they now must almost think the roofs
Threaten to ruin down upon their heads.
And now, when nature begins to lift on high 250
The sun's red splendour and the tremulous fires,
And raise him o'er the mountain-tops, those mountains–
O'er which he seemeth then to thee to be,
His glowing self hard by atingeing them
With his own fire–are yet away from us 255
Scarcely two thousand arrow-shots, indeed
Oft scarce five hundred courses of a dart;
Although between those mountains and the sun
Lie the huge plains of ocean spread beneath
The vasty shores of ether, and intervene 260
A thousand lands, possessed by many a folk
And generations of wild beasts. Again,
A pool of water of but a finger's depth,
Which lies between the stones along the pave,
Offers a vision downward into earth 265
As far, as from the earth o'erspread on high
The gulfs of heaven; that thus thou seemest to view
Clouds down below and heavenly bodies plunged
Wondrously in heaven under earth.
Then too, when in the middle of the stream 270
Sticks fast our dashing horse, and down we gaze
Into the river's rapid waves, some force
Seems then to bear the body of the horse,
Though standing still, reversely from his course,
And swiftly push up-stream. And wheresoe'er 275

We cast our eyes across, all objects seem
Thus to be onward borne and flow along
In the same way as we. A portico,
Albeit it stands well propped from end to end
On equal columns, parallel and big, 280
Contracts by stages in a narrow cone,
When from one end the long, long whole is seen,–
Until, conjoining ceiling with the floor,
And the whole right side with the left, it draws
Together to a cone's nigh-viewless point. 285
To sailors on the main the sun he seems
From out the waves to rise, and in the waves
To set and bury his light–because indeed
They gaze on naught but water and the sky.
Again, to gazers ignorant of the sea, 290
Vessels in port seem, as with broken poops,
To lean upon the water, quite agog;
For any portion of the oars that's raised
Above the briny spray is straight, and straight
The rudders from above. But other parts, 295
Those sunk, immersed below the water-line,
Seem broken all and bended and inclined
Sloping to upwards, and turned back to float
Almost atop the water. And when the winds
Carry the scattered drifts along the sky 300
In the night-time, then seem to glide along
The radiant constellations 'gainst the clouds
And there on high to take far other course
From that whereon in truth they're borne. And then,
If haply our hand be set beneath one eye 305
And press below thereon, then to our gaze
Each object which we gaze on seems to be,
By some sensation twain–then twain the lights
Of lampions burgeoning in flowers of flame,
And twain the furniture in all the house, 310
Two-fold the visages of fellow-men,
And twain their bodies. And again, when sleep
Has bound our members down in slumber soft
And all the body lies in deep repose,
Yet then we seem to self to be awake 315
And move our members; and in night's blind gloom

We think to mark the daylight and the sun;
And, shut within a room, yet still we seem
To change our skies, our oceans, rivers, hills,
To cross the plains afoot, and hear new sounds, 320
Though still the austere silence of the night
Abides around us, and to speak replies,
Though voiceless. Other cases of the sort
Wondrously many do we see, which all
Seek, so to say, to injure faith in sense– 325
In vain, because the largest part of these
Deceives through mere opinions of the mind,
Which we do add ourselves, feigning to see
What by the senses are not seen at all.
For naught is harder than to separate 330
Plain facts from dubious, which the mind forthwith
Adds by itself.

Again, if one suppose
That naught is known, he knows not whether this 335
Itself is able to be known, since he
Confesses naught to know. Therefore with him
I waive discussion–who has set his head
Even where his feet should be. But let me grant
That this he knows,–I question: whence he knows 340
What 'tis to know and not-to-know in turn,
And what created concept of the truth,
And what device has proved the dubious
To differ from the certain?–since in things
He's heretofore seen naught of true. Thou'lt find 345
That from the senses first hath been create
Concept of truth, nor can the senses be
Rebutted. For criterion must be found
Worthy of greater trust, which shall defeat
Through own authority the false by true; 350
What, then, than these our senses must there be
Worthy a greater trust? Shall reason, sprung
From some false sense, prevail to contradict
Those senses, sprung as reason wholly is
From out the senses?–For lest these be true, 355
All reason also then is falsified.
Or shall the ears have power to blame the eyes,

Or yet the touch the ears? Again, shall taste
Accuse this touch or shall the nose confute
Or eyes defeat it? Methinks not so it is: 360
For unto each has been divided off
Its function quite apart, its power to each;
And thus we're still constrained to perceive
The soft, the cold, the hot apart, apart
All divers hues and whatso things there be 365
Conjoined with hues. Likewise the tasting tongue
Has its own power apart, and smells apart
And sounds apart are known. And thus it is
That no one sense can e'er convict another.
Nor shall one sense have power to blame itself, 370
Because it always must be deemed the same,
Worthy of equal trust. And therefore what
At any time unto these senses showed,
The same is true. And if the reason be
Unable to unravel us the cause 375
Why objects, which at hand were square, afar
Seemed rounded, yet it more availeth us,
Lacking the reason, to pretend a cause
For each configuration, than to let
From out our hands escape the obvious things 380
And injure primal faith in sense, and wreck
All those foundations upon which do rest
Our life and safety. For not only reason
Would topple down; but even our very life
Would straightaway collapse, unless we dared 385
To trust our senses and to keep away
From headlong heights and places to be shunned
Of a like peril, and to seek with speed
Their opposites! Again, as in a building,
If the first plumb-line be askew, and if 390
The square deceiving swerve from lines exact,
And if the level waver but the least
In any part, the whole construction then
Must turn out faulty–shelving and askew,
Leaning to back and front, incongruous, 395
That now some portions seem about to fall,
And falls the whole ere long–betrayed indeed
By first deceiving estimates: so too

Thy calculations in affairs of life
Must be askew and false, if sprung for thee 400
From senses false. So all that troop of words
Marshalled against the senses is quite vain.

And now remains to demonstrate with ease
How other senses each their things perceive. 405

Firstly, a sound and every voice is heard,
When, getting into ears, they strike the sense
With their own body. For confess we must
Even voice and sound to be corporeal, 410
Because they're able on the sense to strike.
Besides voice often scrapes against the throat,
And screams in going out do make more rough
The wind-pipe–naturally enough, methinks,
When, through the narrow exit rising up 415
In larger throng, these primal germs of voice
Have thus begun to issue forth. In sooth,
Also the door of the mouth is scraped against
[By air blown outward] from distended [cheeks].

420

And thus no doubt there is, that voice and words
Consist of elements corporeal,
With power to pain. Nor art thou unaware
Likewise how much of body's ta'en away,
How much from very thews and powers of men 425
May be withdrawn by steady talk, prolonged
Even from the rising splendour of the morn
To shadows of black evening,–above all
If 't be outpoured with most exceeding shouts.
Therefore the voice must be corporeal, 430
Since the long talker loses from his frame
A part.

Moreover, roughness in the sound
Comes from the roughness in the primal germs, 435
As a smooth sound from smooth ones is create;
Nor have these elements a form the same
When the trump rumbles with a hollow roar,
As when barbaric Berecynthian pipe

Buzzes with raucous boomings, or when swans 440
By night from icy shores of Helicon
With wailing voices raise their liquid dirge.

Thus, when from deep within our frame we force
These voices, and at mouth expel them forth, 445
The mobile tongue, artificer of words,
Makes them articulate, and too the lips
By their formations share in shaping them.
Hence when the space is short from starting-point
To where that voice arrives, the very words 450
Must too be plainly heard, distinctly marked.
For then the voice conserves its own formation,
Conserves its shape. But if the space between
Be longer than is fit, the words must be
Through the much air confounded, and the voice 455
Disordered in its flight across the winds–
And so it haps, that thou canst sound perceive,
Yet not determine what the words may mean;
To such degree confounded and encumbered
The voice approaches us. Again, one word, 460
Sent from the crier's mouth, may rouse all ears
Among the populace. And thus one voice
Scatters asunder into many voices,
Since it divides itself for separate ears,
Imprinting form of word and a clear tone. 465
But whatso part of voices fails to hit
The ears themselves perishes, borne beyond,
Idly diffused among the winds. A part,
Beating on solid porticoes, tossed back
Returns a sound; and sometimes mocks the ear 470
With a mere phantom of a word. When this
Thou well hast noted, thou canst render count
Unto thyself and others why it is
Along the lonely places that the rocks
Give back like shapes of words in order like, 475
When search we after comrades wandering
Among the shady mountains, and aloud
Call unto them, the scattered. I have seen
Spots that gave back even voices six or seven
For one thrown forth–for so the very hills, 480

Dashing them back against the hills, kept on
With their reverberations. And these spots
The neighbouring country-side doth feign to be
Haunts of the goat-foot satyrs and the nymphs;
And tells ye there be fauns, by whose night noise 485
And antic revels yonder they declare
The voiceless silences are broken oft,
And tones of strings are made and wailings sweet
Which the pipe, beat by players' finger-tips,
Pours out; and far and wide the farmer-race 490
Begins to hear, when, shaking the garmentings
Of pine upon his half-beast head, god-Pan
With puckered lip oft runneth o'er and o'er
The open reeds,–lest flute should cease to pour
The woodland music! Other prodigies 495
And wonders of this ilk they love to tell,
Lest they be thought to dwell in lonely spots
And even by gods deserted. This is why
They boast of marvels in their story-tellings;
Or by some other reason are led on– 500
Greedy, as all mankind hath ever been,
To prattle fables into ears.

Again,
One need not wonder how it comes about 505
That through those places (through which eyes cannot
View objects manifest) sounds yet may pass
And assail the ears. For often we observe
People conversing, though the doors be closed;
No marvel either, since all voice unharmed 510
Can wind through bended apertures of things,
While idol-films decline to–for they're rent,
Unless along straight apertures they swim,
Like those in glass, through which all images
Do fly across. And yet this voice itself, 515
In passing through shut chambers of a house,
Is dulled, and in a jumble enters ears,
And sound we seem to hear far more than words.
Moreover, a voice is into all directions
Divided up, since off from one another 520
New voices are engendered, when one voice

Hath once leapt forth, outstarting into many–
As oft a spark of fire is wont to sprinkle
Itself into its several fires. And so,
Voices do fill those places hid behind, 525
Which all are in a hubbub round about,
Astir with sound. But idol-films do tend,
As once sent forth, in straight directions all;
Wherefore one can inside a wall see naught,
Yet catch the voices from beyond the same. 530

Nor tongue and palate, whereby we flavour feel,
Present more problems for more work of thought.
Firstly, we feel a flavour in the mouth,
When forth we squeeze it, in chewing up our food,– 535
As any one perchance begins to squeeze
With hand and dry a sponge with water soaked.
Next, all which forth we squeeze is spread about
Along the pores and intertwined paths
Of the loose-textured tongue. And so, when smooth 540
The bodies of the oozy flavour, then
Delightfully they touch, delightfully
They treat all spots, around the wet and trickling
Enclosures of the tongue. And contrariwise,
They sting and pain the sense with their assault, 545
According as with roughness they're supplied.
Next, only up to palate is the pleasure
Coming from flavour; for in truth when down
'Thas plunged along the throat, no pleasure is,
Whilst into all the frame it spreads around; 550
Nor aught it matters with what food is fed
The body, if only what thou take thou canst
Distribute well digested to the frame
And keep the stomach in a moist career.
555

Now, how it is we see some food for some,
Others for others....

I will unfold, or wherefore what to some
Is foul and bitter, yet the same to others 560
Can seem delectable to eat,–why here
So great the distance and the difference is

That what is food to one to some becomes
Fierce poison, as a certain snake there is
Which, touched by spittle of a man, will waste 565
And end itself by gnawing up its coil.
Again, fierce poison is the hellebore
To us, but puts the fat on goats and quails.
That thou mayst know by what devices this
Is brought about, in chief thou must recall 570
What we have said before, that seeds are kept
Commixed in things in divers modes. Again,
As all the breathing creatures which take food
Are outwardly unlike, and outer cut
And contour of their members bounds them round, 575
Each differing kind by kind, they thus consist
Of seeds of varying shape. And furthermore,
Since seeds do differ, divers too must be
The interstices and paths (which we do call
The apertures) in all the members, even 580
In mouth and palate too. Thus some must be
More small or yet more large, three-cornered some
And others squared, and many others round,
And certain of them many-angled too
In many modes. For, as the combination 585
And motion of their divers shapes demand,
The shapes of apertures must be diverse
And paths must vary according to their walls
That bound them. Hence when what is sweet to some,
Becomes to others bitter, for him to whom 590
'Tis sweet, the smoothest particles must needs
Have entered caressingly the palate's pores.
And, contrariwise, with those to whom that sweet
Is sour within the mouth, beyond a doubt
The rough and barbed particles have got 595
Into the narrows of the apertures.
Now easy it is from these affairs to know
Whatever...

Indeed, where one from o'er-abundant bile 600
Is stricken with fever, or in other wise
Feels the roused violence of some malady,
There the whole frame is now upset, and there

All the positions of the seeds are changed,–
So that the bodies which before were fit 605
To cause the savour, now are fit no more,
And now more apt are others which be able
To get within the pores and gender sour.
Both sorts, in sooth, are intermixed in honey–
What oft we've proved above to thee before. 610
Now come, and I will indicate what wise
Impact of odour on the nostrils touches.
And first, 'tis needful there be many things
From whence the streaming flow of varied odours
May roll along, and we're constrained to think 615
They stream and dart and sprinkle themselves about
Impartially. But for some breathing creatures
One odour is more apt, to others another–
Because of differing forms of seeds and pores.
Thus on and on along the zephyrs bees 620
Are led by odour of honey, vultures too
By carcasses. Again, the forward power
Of scent in dogs doth lead the hunter on
Whithersoever the splay-foot of wild beast
Hath hastened its career; and the white goose, 625
The saviour of the Roman citadel,
Forescents afar the odour of mankind.
Thus, diversly to divers ones is given
Peculiar smell that leadeth each along
To his own food or makes him start aback 630
From loathsome poison, and in this wise are
The generations of the wild preserved.

Yet is this pungence not alone in odours
Or in the class of flavours; but, likewise, 635
The look of things and hues agree not all
So well with senses unto all, but that
Some unto some will be, to gaze upon,
More keen and painful. Lo, the raving lions,
They dare not face and gaze upon the cock 640
Who's wont with wings to flap away the night
From off the stage, and call the beaming morn
With clarion voice–and lions straightway thus
Bethink themselves of flight, because, ye see,

Within the body of the cocks there be 645
Some certain seeds, which, into lions' eyes
Injected, bore into the pupils deep
And yield such piercing pain they can't hold out
Against the cocks, however fierce they be–
Whilst yet these seeds can't hurt our gaze the least, 650
Either because they do not penetrate,
Or since they have free exit from the eyes
As soon as penetrating, so that thus
They cannot hurt our eyes in any part
By there remaining. 655

To speak once more of odour;
Whatever assail the nostrils, some can travel
A longer way than others. None of them,
However, 's borne so far as sound or voice– 660
While I omit all mention of such things
As hit the eyesight and assail the vision.
For slowly on a wandering course it comes
And perishes sooner, by degrees absorbed
Easily into all the winds of air;– 665
And first, because from deep inside the thing
It is discharged with labour (for the fact
That every object, when 'tis shivered, ground,
Or crumbled by the fire, will smell the stronger
Is sign that odours flow and part away 670
From inner regions of the things). And next,
Thou mayest see that odour is create
Of larger primal germs than voice, because
It enters not through stony walls, wherethrough
Unfailingly the voice and sound are borne; 675
Wherefore, besides, thou wilt observe 'tis not
So easy to trace out in whatso place
The smelling object is. For, dallying on
Along the winds, the particles cool off,
And then the scurrying messengers of things 680
Arrive our senses, when no longer hot.
So dogs oft wander astray, and hunt the scent.

Now mark, and hear what objects move the mind,
And learn, in few, whence unto intellect 685

Do come what come. And first I tell thee this:
That many images of objects rove
In many modes to every region round–
So thin that easily the one with other,
When once they meet, uniteth in mid-air, 690
Like gossamer or gold-leaf. For, indeed,
Far thinner are they in their fabric than
Those images which take a hold on eyes
And smite the vision, since through body's pores
They penetrate, and inwardly stir up 695
The subtle nature of mind and smite the sense.
Thus, Centaurs and the limbs of Scyllas, thus
The Cerberus-visages of dogs we see,
And images of people gone before–
Dead men whose bones earth bosomed long ago; 700
Because the images of every kind
Are everywhere about us borne–in part
Those which are gendered in the very air
Of own accord, in part those others which
From divers things do part away, and those 705
Which are compounded, made from out their shapes.
For soothly from no living Centaur is
That phantom gendered, since no breed of beast
Like him was ever; but, when images
Of horse and man by chance have come together, 710
They easily cohere, as aforesaid,
At once, through subtle nature and fabric thin.
In the same fashion others of this ilk
Created are. And when they're quickly borne
In their exceeding lightness, easily 715
(As earlier I showed) one subtle image,
Compounded, moves by its one blow the mind,
Itself so subtle and so strangely quick.

That these things come to pass as I record, 720
From this thou easily canst understand:
So far as one is unto other like,
Seeing with mind as well as with the eyes
Must come to pass in fashion not unlike.
Well, now, since I have shown that I perceive 725
Haply a lion through those idol-films

Such as assail my eyes, 'tis thine to know
Also the mind is in like manner moved,
And sees, nor more nor less than eyes do see
(Except that it perceives more subtle films) 730
The lion and aught else through idol-films.
And when the sleep has overset our frame,
The mind's intelligence is now awake,
Still for no other reason, save that these–
The self-same films as when we are awake– 735
Assail our minds, to such degree indeed
That we do seem to see for sure the man
Whom, void of life, now death and earth have gained
Dominion over. And nature forces this
To come to pass because the body's senses 740
Are resting, thwarted through the members all,
Unable now to conquer false with true;
And memory lies prone and languishes
In slumber, nor protests that he, the man
Whom the mind feigns to see alive, long since 745
Hath been the gain of death and dissolution.

And further, 'tis no marvel idols move
And toss their arms and other members round
In rhythmic time–and often in men's sleeps 750
It haps an image this is seen to do;
In sooth, when perishes the former image,
And other is gendered of another pose,
That former seemeth to have changed its gestures.
Of course the change must be conceived as speedy; 755
So great the swiftness and so great the store
Of idol-things, and (in an instant brief
As mind can mark) so great, again, the store
Of separate idol-parts to bring supplies.

760

It happens also that there is supplied
Sometimes an image not of kind the same;
But what before was woman, now at hand
Is seen to stand there, altered into male;
Or other visage, other age succeeds; 765
But slumber and oblivion take care
That we shall feel no wonder at the thing.

And much in these affairs demands inquiry,
And much, illumination–if we crave
With plainness to exhibit facts. And first, 770
Why doth the mind of one to whom the whim
To think has come behold forthwith that thing?
Or do the idols watch upon our will,
And doth an image unto us occur,
Directly we desire–if heart prefer 775
The sea, the land, or after all the sky?
Assemblies of the citizens, parades,
Banquets, and battles, these and all doth she,
Nature, create and furnish at our word?–
Maugre the fact that in same place and spot 780
Another's mind is meditating things
All far unlike. And what, again, of this:
When we in sleep behold the idols step,
In measure, forward, moving supple limbs,
Whilst forth they put each supple arm in turn 785
With speedy motion, and with eyeing heads
Repeat the movement, as the foot keeps time?
Forsooth, the idols they are steeped in art,
And wander to and fro well taught indeed,–
Thus to be able in the time of night 790
To make such games! Or will the truth be this:
Because in one least moment that we mark–
That is, the uttering of a single sound–
There lurk yet many moments, which the reason
Discovers to exist, therefore it comes 795
That, in a moment how so brief ye will,
The divers idols are hard by, and ready
Each in its place diverse? So great the swiftness,
So great, again, the store of idol-things,
And so, when perishes the former image, 800
And other is gendered of another pose,
The former seemeth to have changed its gestures.
And since they be so tenuous, mind can mark
Sharply alone the ones it strains to see;
And thus the rest do perish one and all, 805
Save those for which the mind prepares itself.
Further, it doth prepare itself indeed,
And hopes to see what follows after each–

Hence this result. For hast thou not observed
How eyes, essaying to perceive the fine, 810
Will strain in preparation, otherwise
Unable sharply to perceive at all?
Yet know thou canst that, even in objects plain,
If thou attendest not, 'tis just the same
As if 'twere all the time removed and far. 815
What marvel, then, that mind doth lose the rest,
Save those to which 'thas given up itself?
So 'tis that we conjecture from small signs
Things wide and weighty, and involve ourselves
In snarls of self-deceit. 820

Some Vital Functions

In these affairs
We crave that thou wilt passionately flee
The one offence, and anxiously wilt shun
The error of presuming the clear lights
Of eyes created were that we might see; 5
Or thighs and knees, aprop upon the feet,
Thuswise can bended be, that we might step
With goodly strides ahead; or forearms joined
Unto the sturdy uppers, or serving hands
On either side were given, that we might do 10
Life's own demands. All such interpretation
Is aft-for-fore with inverse reasoning,
Since naught is born in body so that we
May use the same, but birth engenders use:
No seeing ere the lights of eyes were born, 15
No speaking ere the tongue created was;
But origin of tongue came long before
Discourse of words, and ears created were
Much earlier than any sound was heard;
And all the members, so meseems, were there 20
Before they got their use: and therefore, they
Could not be gendered for the sake of use.
But contrariwise, contending in the fight
With hand to hand, and rending of the joints,
And fouling of the limbs with gore, was there, 25
O long before the gleaming spears ere flew;
And nature prompted man to shun a wound,
Before the left arm by the aid of art
Opposed the shielding targe. And, verily,

Yielding the weary body to repose, 30
Far ancienter than cushions of soft beds,
And quenching thirst is earlier than cups.
These objects, therefore, which for use and life
Have been devised, can be conceived as found
For sake of using. But apart from such 35
Are all which first were born and afterwards
Gave knowledge of their own utility–
Chief in which sort we note the senses, limbs:
Wherefore, again, 'tis quite beyond thy power
To hold that these could thus have been create 40
For office of utility.

Likewise,
'Tis nothing strange that all the breathing creatures
Seek, even by nature of their frame, their food. 45
Yes, since I've taught thee that from off the things
Stream and depart innumerable bodies
In modes innumerable too; but most
Must be the bodies streaming from the living–
Which bodies, vexed by motion evermore, 50
Are through the mouth exhaled innumerable,
When weary creatures pant, or through the sweat
Squeezed forth innumerable from deep within.
Thus body rarefies, so undermined
In all its nature, and pain attends its state. 55
And so the food is taken to underprop
The tottering joints, and by its interfusion
To re-create their powers, and there stop up
The longing, open-mouthed through limbs and veins,
For eating. And the moist no less departs 60
Into all regions that demand the moist;
And many heaped-up particles of hot,
Which cause such burnings in these bellies of ours,
The liquid on arriving dissipates
And quenches like a fire, that parching heat 65
No longer now can scorch the frame. And so,
Thou seest how panting thirst is washed away
From off our body, how the hunger-pang
It, too, appeased.

70

Now, how it comes that we,
Whene'er we wish, can step with strides ahead,
And how 'tis given to move our limbs about,
And what device is wont to push ahead
This the big load of our corporeal frame, 75
I'll say to thee–do thou attend what's said.
I say that first some idol-films of walking
Into our mind do fall and smite the mind,
As said before. Thereafter will arises;
For no one starts to do a thing, before 80
The intellect previsions what it wills;
And what it there pre-visioneth depends
On what that image is. When, therefore, mind
Doth so bestir itself that it doth will
To go and step along, it strikes at once 85
That energy of soul that's sown about
In all the body through the limbs and frame–
And this is easy of performance, since
The soul is close conjoined with the mind.
Next, soul in turn strikes body, and by degrees 90
Thus the whole mass is pushed along and moved.
Then too the body rarefies, and air,
Forsooth as ever of such nimbleness,
Comes on and penetrates aboundingly
Through opened pores, and thus is sprinkled round 95
Unto all smallest places in our frame.
Thus then by these twain factors, severally,
Body is borne like ship with oars and wind.
Nor yet in these affairs is aught for wonder
That particles so fine can whirl around 100
So great a body and turn this weight of ours;
For wind, so tenuous with its subtle body,
Yet pushes, driving on the mighty ship
Of mighty bulk; one hand directs the same,
Whatever its momentum, and one helm 105
Whirls it around, whither ye please; and loads,
Many and huge, are moved and hoisted high
By enginery of pulley-blocks and wheels,
With but light strain.

110

Now, by what modes this sleep

Pours through our members waters of repose
And frees the breast from cares of mind, I'll tell
In verses sweeter than they many are;
Even as the swan's slight note is better far 115
Than that dispersed clamour of the cranes
Among the southwind's aery clouds. Do thou
Give me sharp ears and a sagacious mind,–
That thou mayst not deny the things to be
Whereof I'm speaking, nor depart away 120
With bosom scorning these the spoken truths,
Thyself at fault unable to perceive.
Sleep chiefly comes when energy of soul
Hath now been scattered through the frame, and part
Expelled abroad and gone away, and part 125
Crammed back and settling deep within the frame–
Whereafter then our loosened members droop.
For doubt is none that by the work of soul
Exist in us this sense, and when by slumber
That sense is thwarted, we are bound to think 130
The soul confounded and expelled abroad–
Yet not entirely, else the frame would lie
Drenched in the everlasting cold of death.
In sooth, where no one part of soul remained
Lurking among the members, even as fire 135
Lurks buried under many ashes, whence
Could sense amain rekindled be in members,
As flame can rise anew from unseen fire?

By what devices this strange state and new 140
May be occasioned, and by what the soul
Can be confounded and the frame grow faint,
I will untangle: see to it, thou, that I
Pour forth my words not unto empty winds.
In first place, body on its outer parts– 145
Since these are touched by neighbouring aery gusts–
Must there be thumped and strook by blows of air
Repeatedly. And therefore almost all
Are covered either with hides, or else with shells,
Or with the horny callus, or with bark. 150
Yet this same air lashes their inner parts,
When creatures draw a breath or blow it out.

Wherefore, since body thus is flogged alike
Upon the inside and the out, and blows
Come in upon us through the little pores 155
Even inward to our body's primal parts
And primal elements, there comes to pass
By slow degrees, along our members then,
A kind of overthrow; for then confounded
Are those arrangements of the primal germs 160
Of body and of mind. It comes to pass
That next a part of soul's expelled abroad,
A part retreateth in recesses hid,
A part, too, scattered all about the frame,
Cannot become united nor engage 165
In interchange of motion. Nature now
So hedges off approaches and the paths;
And thus the sense, its motions all deranged,
Retires down deep within; and since there's naught,

170

As 'twere, to prop the frame, the body weakens,
And all the members languish, and the arms
And eyelids fall, and, as ye lie abed,
Even there the houghs will sag and loose their powers.
Again, sleep follows after food, because 175
The food produces same result as air,
Whilst being scattered round through all the veins;
And much the heaviest is that slumber which,
Full or fatigued, thou takest; since 'tis then
That the most bodies disarrange themselves, 180
Bruised by labours hard. And in same wise,
This three-fold change: a forcing of the soul
Down deeper, more a casting-forth of it,
A moving more divided in its parts
And scattered more. 185

And to whate'er pursuit
A man most clings absorbed, or what the affairs
On which we theretofore have tarried much,
And mind hath strained upon the more, we seem 190
In sleep not rarely to go at the same.
The lawyers seem to plead and cite decrees,
Commanders they to fight and go at frays,

Sailors to live in combat with the winds,
And we ourselves indeed to make this book, 195
And still to seek the nature of the world
And set it down, when once discovered, here
In these my country's leaves. Thus all pursuits,
All arts in general seem in sleeps to mock
And master the minds of men. And whosoever 200
Day after day for long to games have given
Attention undivided, still they keep
(As oft we note), even when they've ceased to grasp
Those games with their own senses, open paths
Within the mind wherethrough the idol-films 205
Of just those games can come. And thus it is
For many a day thereafter those appear
Floating before the eyes, that even awake
They think they view the dancers moving round
Their supple limbs, and catch with both the ears 210
The liquid song of harp and speaking chords,
And view the same assembly on the seats,
And manifold bright glories of the stage–
So great the influence of pursuit and zest,
And of the affairs wherein 'thas been the wont 215
Of men to be engaged-nor only men,
But soothly all the animals. Behold,
Thou'lt see the sturdy horses, though outstretched,
Yet sweating in their sleep, and panting ever,
And straining utmost strength, as if for prize, 220
As if, with barriers opened now...
And hounds of huntsmen oft in soft repose
Yet toss asudden all their legs about,
And growl and bark, and with their nostrils sniff
The winds again, again, as though indeed 225
They'd caught the scented foot-prints of wild beasts,
And, even when wakened, often they pursue
The phantom images of stags, as though
They did perceive them fleeing on before,
Until the illusion's shaken off and dogs 230
Come to themselves again. And fawning breed
Of house-bred whelps do feel the sudden urge
To shake their bodies and start from off the ground,
As if beholding stranger-visages.

And ever the fiercer be the stock, the more 235
In sleep the same is ever bound to rage.
But flee the divers tribes of birds and vex
With sudden wings by night the groves of gods,
When in their gentle slumbers they have dreamed
Of hawks in chase, aswooping on for fight. 240
Again, the minds of mortals which perform
With mighty motions mighty enterprises,
Often in sleep will do and dare the same
In manner like. Kings take the towns by storm,
Succumb to capture, battle on the field, 245
Raise a wild cry as if their throats were cut
Even then and there. And many wrestle on
And groan with pains, and fill all regions round
With mighty cries and wild, as if then gnawed
By fangs of panther or of lion fierce. 250
Many amid their slumbers talk about
Their mighty enterprises, and have often
Enough become the proof of their own crimes.
Many meet death; many, as if headlong
From lofty mountains tumbling down to earth 255
With all their frame, are frenzied in their fright;
And after sleep, as if still mad in mind,
They scarce come to, confounded as they are
By ferment of their frame. The thirsty man,
Likewise, he sits beside delightful spring 260
Or river and gulpeth down with gaping throat
Nigh the whole stream. And oft the innocent young,
By sleep o'ermastered, think they lift their dress
By pail or public jordan and then void
The water filtered down their frame entire 265
And drench the Babylonian coverlets,
Magnificently bright. Again, those males
Into the surging channels of whose years
Now first has passed the seed (engendered
Within their members by the ripened days) 270
Are in their sleep confronted from without
By idol-images of some fair form–
Tidings of glorious face and lovely bloom,
Which stir and goad the regions turgid now
With seed abundant; so that, as it were 275

With all the matter acted duly out,
They pour the billows of a potent stream
And stain their garment.

And as said before, 280
That seed is roused in us when once ripe age
Has made our body strong...
As divers causes give to divers things
Impulse and irritation, so one force
In human kind rouses the human seed 285
To spurt from man. As soon as ever it issues,
Forced from its first abodes, it passes down
In the whole body through the limbs and frame,
Meeting in certain regions of our thews,
And stirs amain the genitals of man. 290
The goaded regions swell with seed, and then
Comes the delight to dart the same at what
The mad desire so yearns, and body seeks
That object, whence the mind by love is pierced.
For well-nigh each man falleth toward his wound, 295
And our blood spurts even toward the spot from whence
The stroke wherewith we are strook, and if indeed
The foe be close, the red jet reaches him.
Thus, one who gets a stroke from Venus' shafts–
Whether a boy with limbs effeminate 300
Assault him, or a woman darting love
From all her body–that one strains to get
Even to the thing whereby he's hit, and longs
To join with it and cast into its frame
The fluid drawn even from within its own. 305
For the mute craving doth presage delight.

The Passion of Love

This craving 'tis that's Venus unto us:
From this, engender all the lures of love,
From this, O first hath into human hearts
Trickled that drop of joyance which ere long
Is by chill care succeeded. Since, indeed, 5
Though she thou lovest now be far away,
Yet idol-images of her are near
And the sweet name is floating in thy ear.
But it behooves to flee those images;
And scare afar whatever feeds thy love; 10
And turn elsewhere thy mind; and vent the sperm,
Within thee gathered, into sundry bodies,
Nor, with thy thoughts still busied with one love,
Keep it for one delight, and so store up
Care for thyself and pain inevitable. 15
For, lo, the ulcer just by nourishing
Grows to more life with deep inveteracy,
And day by day the fury swells aflame,
And the woe waxes heavier day by day–
Unless thou dost destroy even by new blows 20
The former wounds of love, and curest them
While yet they're fresh, by wandering freely round
After the freely-wandering Venus, or
Canst lead elsewhere the tumults of thy mind.
25

Nor doth that man who keeps away from love
Yet lack the fruits of Venus; rather takes
Those pleasures which are free of penalties.
For the delights of Venus, verily,

Are more unmixed for mortals sane-of-soul 30
Than for those sick-at-heart with love-pining.
Yea, in the very moment of possessing,
Surges the heat of lovers to and fro,
Restive, uncertain; and they cannot fix
On what to first enjoy with eyes and hands. 35
The parts they sought for, those they squeeze so tight,
And pain the creature's body, close their teeth
Often against her lips, and smite with kiss
Mouth into mouth,–because this same delight
Is not unmixed; and underneath are stings 40
Which goad a man to hurt the very thing,
Whate'er it be, from whence arise for him
Those germs of madness. But with gentle touch
Venus subdues the pangs in midst of love,
And the admixture of a fondling joy 45
Doth curb the bites of passion. For they hope
That by the very body whence they caught
The heats of love their flames can be put out.
But nature protests 'tis all quite otherwise;
For this same love it is the one sole thing 50
Of which, the more we have, the fiercer burns
The breast with fell desire. For food and drink
Are taken within our members; and, since they
Can stop up certain parts, thus, easily
Desire of water is glutted and of bread. 55
But, lo, from human face and lovely bloom
Naught penetrates our frame to be enjoyed
Save flimsy idol-images and vain–
A sorry hope which oft the winds disperse.
As when the thirsty man in slumber seeks 60
To drink, and water ne'er is granted him
Wherewith to quench the heat within his members,
But after idols of the liquids strives
And toils in vain, and thirsts even whilst he gulps
In middle of the torrent, thus in love 65
Venus deludes with idol-images
The lovers. Nor they cannot sate their lust
By merely gazing on the bodies, nor
They cannot with their palms and fingers rub
Aught from each tender limb, the while they stray 70

Uncertain over all the body. Then,
At last, with members intertwined, when they
Enjoy the flower of their age, when now
Their bodies have sweet presage of keen joys,
And Venus is about to sow the fields 75
Of woman, greedily their frames they lock,
And mingle the slaver of their mouths, and breathe
Into each other, pressing teeth on mouths–
Yet to no purpose, since they're powerless
To rub off aught, or penetrate and pass 80
With body entire into body–for oft
They seem to strive and struggle thus to do;
So eagerly they cling in Venus' bonds,
Whilst melt away their members, overcome
By violence of delight. But when at last 85
Lust, gathered in the thews, hath spent itself,
There come a brief pause in the raging heat–
But then a madness just the same returns
And that old fury visits them again,
When once again they seek and crave to reach 90
They know not what, all powerless to find
The artifice to subjugate the bane.
In such uncertain state they waste away
With unseen wound.

95

To which be added too,
They squander powers and with the travail wane;
Be added too, they spend their futile years
Under another's beck and call; their duties
Neglected languish and their honest name 100
Reeleth sick, sick; and meantime their estates
Are lost in Babylonian tapestries;
And unguents and dainty Sicyonian shoes
Laugh on her feet; and (as ye may be sure)
Big emeralds of green light are set in gold; 105
And rich sea-purple dress by constant wear
Grows shabby and all soaked with Venus' sweat;
And the well-earned ancestral property
Becometh head-bands, coifs, and many a time
The cloaks, or garments Alidensian 110
Or of the Cean isle. And banquets, set

With rarest cloth and viands, are prepared–
And games of chance, and many a drinking cup,
And unguents, crowns and garlands. All in vain,
Since from amid the well-spring of delights 115
Bubbles some drop of bitter to torment
Among the very flowers–when haply mind
Gnaws into self, now stricken with remorse
For slothful years and ruin in baudels,
Or else because she's left him all in doubt 120
By launching some sly word, which still like fire
Lives wildly, cleaving to his eager heart;
Or else because he thinks she darts her eyes
Too much about and gazes at another,–
And in her face sees traces of a laugh. 125

These ills are found in prospering love and true;
But in crossed love and helpless there be such
As through shut eyelids thou canst still take in–
Uncounted ills; so that 'tis better far 130
To watch beforehand, in the way I've shown,
And guard against enticements. For to shun
A fall into the hunting-snares of love
Is not so hard, as to get out again,
When tangled in the very nets, and burst 135
The stoutly-knotted cords of Aphrodite.
Yet even when there enmeshed with tangled feet,
Still canst thou scape the danger-lest indeed
Thou standest in the way of thine own good,
And overlookest first all blemishes 140
Of mind and body of thy much preferred,
Desirable dame. For so men do,
Eyeless with passion, and assign to them
Graces not theirs in fact. And thus we see
Creatures in many a wise crooked and ugly 145
The prosperous sweethearts in a high esteem;
And lovers gird each other and advise
To placate Venus, since their friends are smit
With a base passion–miserable dupes
Who seldom mark their own worst bane of all. 150
The black-skinned girl is "tawny like the honey";
The filthy and the fetid's "negligee";

The cat-eyed she's "a little Pallas," she;
The sinewy and wizened's "a gazelle";
The pudgy and the pigmy is "piquant, 155
One of the Graces sure"; the big and bulky
O she's "an Admiration, imposante";
The stuttering and tongue-tied "sweetly lisps";
The mute girl's "modest"; and the garrulous,
The spiteful spit-fire, is "a sparkling wit"; 160
And she who scarcely lives for scrawniness
Becomes "a slender darling"; "delicate"
Is she who's nearly dead of coughing-fit;
The pursy female with protuberant breasts
She is "like Ceres when the goddess gave 165
Young Bacchus suck"; the pug-nosed lady-love
"A Satyress, a feminine Silenus";
The blubber-lipped is "all one luscious kiss"–
A weary while it were to tell the whole.
But let her face possess what charm ye will, 170
Let Venus' glory rise from all her limbs,–
Forsooth there still are others; and forsooth
We lived before without her; and forsooth
She does the same things–and we know she does–
All, as the ugly creature, and she scents, 175
Yes she, her wretched self with vile perfumes;
Whom even her handmaids flee and giggle at
Behind her back. But he, the lover, in tears
Because shut out, covers her threshold o'er
Often with flowers and garlands, and anoints 180
Her haughty door-posts with the marjoram,
And prints, poor fellow, kisses on the doors–
Admitted at last, if haply but one whiff
Got to him on approaching, he would seek
Decent excuses to go out forthwith; 185
And his lament, long pondered, then would fall
Down at his heels; and there he'd damn himself
For his fatuity, observing how
He had assigned to that same lady more–
Than it is proper to concede to mortals. 190
And these our Venuses are 'ware of this.
Wherefore the more are they at pains to hide
All the-behind-the-scenes of life from those

Whom they desire to keep in bonds of love–
In vain, since ne'ertheless thou canst by thought 195
Drag all the matter forth into the light
And well search out the cause of all these smiles;
And if of graceful mind she be and kind,
Do thou, in thy turn, overlook the same,
And thus allow for poor mortality. 200
Nor sighs the woman always with feigned love,
Who links her body round man's body locked
And holds him fast, making his kisses wet
With lips sucked into lips; for oft she acts
Even from desire, and, seeking mutual joys, 205
Incites him there to run love's race-course through.
Nor otherwise can cattle, birds, wild beasts,
And sheep and mares submit unto the males,
Except that their own nature is in heat,
And burns abounding and with gladness takes 210
Once more the Venus of the mounting males.
And seest thou not how those whom mutual pleasure
Hath bound are tortured in their common bonds?
How often in the cross-roads dogs that pant
To get apart strain eagerly asunder 215
With utmost might?–When all the while they're fast
In the stout links of Venus. But they'd ne'er
So pull, except they knew those mutual joys–
So powerful to cast them unto snares
And hold them bound. Wherefore again, again, 220
Even as I say, there is a joint delight.

And when perchance, in mingling seed with his,
The female hath o'erpowered the force of male
And by a sudden fling hath seized it fast, 225
Then are the offspring, more from mothers' seed,
More like their mothers; as, from fathers' seed,
They're like to fathers. But whom seest to be
Partakers of each shape, one equal blend
Of parents' features, these are generate 230
From fathers' body and from mothers' blood,
When mutual and harmonious heat hath dashed
Together seeds, aroused along their frames
By Venus' goads, and neither of the twain

Mastereth or is mastered. Happens too 235
That sometimes offspring can to being come
In likeness of their grandsires, and bring back
Often the shapes of grandsires' sires, because
Their parents in their bodies oft retain
Concealed many primal germs, commixed 240
In many modes, which, starting with the stock,
Sire handeth down to son, himself a sire;
Whence Venus by a variable chance
Engenders shapes, and diversely brings back
Ancestral features, voices too, and hair. 245
A female generation rises forth
From seed paternal, and from mother's body
Exist created males: since sex proceeds
No more from singleness of seed than faces
Or bodies or limbs of ours: for every birth 250
Is from a twofold seed; and what's created
Hath, of that parent which it is more like,
More than its equal share; as thou canst mark,–
Whether the breed be male or female stock.
255

Nor do the powers divine grudge any man
The fruits of his seed-sowing, so that never
He be called "father" by sweet children his,
And end his days in sterile love forever.
What many men suppose; and gloomily 260
They sprinkle the altars with abundant blood,
And make the high platforms odorous with burnt gifts,
To render big by plenteous seed their wives–
And plague in vain godheads and sacred lots.
For sterile are these men by seed too thick, 265
Or else by far too watery and thin.
Because the thin is powerless to cleave
Fast to the proper places, straightaway
It trickles from them, and, returned again,
Retires abortively. And then since seed 270
More gross and solid than will suit is spent
By some men, either it flies not forth amain
With spurt prolonged enough, or else it fails
To enter suitably the proper places,
Or, having entered, the seed is weakly mixed 275

With seed of the woman: harmonies of Venus
Are seen to matter vastly here; and some
Impregnate some more readily, and from some
Some women conceive more readily and become
Pregnant. And many women, sterile before 280
In several marriage-beds, have yet thereafter
Obtained the mates from whom they could conceive
The baby-boys, and with sweet progeny
Grow rich. And even for husbands (whose own wives,
Although of fertile wombs, have borne for them 285
No babies in the house) are also found
Concordant natures so that they at last
Can bulwark their old age with goodly sons.
A matter of great moment 'tis in truth,
That seeds may mingle readily with seeds 290
Suited for procreation, and that thick
Should mix with fluid seeds, with thick the fluid.
And in this business 'tis of some import
Upon what diet life is nourished:
For some foods thicken seeds within our members, 295
And others thin them out and waste away.
And in what modes the fond delight itself
Is carried on–this too importeth vastly.
For commonly 'tis thought that wives conceive
More readily in manner of wild-beasts, 300
After the custom of the four-foot breeds,
Because so postured, with the breasts beneath
And buttocks then upreared, the seeds can take
Their proper places. Nor is need the least
For wives to use the motions of blandishment; 305
For thus the woman hinders and resists
Her own conception, if too joyously
Herself she treats the Venus of the man
With haunches heaving, and with all her bosom
Now yielding like the billows of the sea– 310
Aye, from the ploughshare's even course and track
She throws the furrow, and from proper places
Deflects the spurt of seed. And courtesans
Are thuswise wont to move for their own ends,
To keep from pregnancy and lying in, 315
And all the while to render Venus more

A pleasure for the men–the which meseems
Our wives have never need of.

Sometimes too 320
It happens–and through no divinity
Nor arrows of Venus–that a sorry chit
Of scanty grace will be beloved by man;
For sometimes she herself by very deeds,
By her complying ways, and tidy habits, 325
Will easily accustom thee to pass
With her thy life-time–and, moreover, lo,
Long habitude can gender human love,
Even as an object smitten o'er and o'er
By blows, however lightly, yet at last 330
Is overcome and wavers. Seest thou not,
Besides, how drops of water falling down
Against the stones at last bore through the stones?

Book Five

Proem

O Who can build with puissant breast a song
Worthy the majesty of these great finds?
Or who in words so strong that he can frame
The fit laudations for deserts of him
Who left us heritors of such vast prizes, 5
By his own breast discovered and sought out?–
There shall be none, methinks, of mortal stock.
For if must needs be named for him the name
Demanded by the now known majesty
Of these high matters, then a god was he,– 10
Hear me, illustrious Memmius–a god;
Who first and chief found out that plan of life
Which now is called philosophy, and who
By cunning craft, out of such mighty waves,
Out of such mighty darkness, moored life 15
In havens so serene, in light so clear.
Compare those old discoveries divine
Of others: lo, according to the tale,
Ceres established for mortality
The grain, and Bacchus juice of vine-born grape, 20
Though life might yet without these things abide,
Even as report saith now some peoples live.
But man's well-being was impossible
Without a breast all free. Wherefore the more
That man doth justly seem to us a god, 25
From whom sweet solaces of life, afar

Distributed o'er populous domains,
Now soothe the minds of men. But if thou thinkest
Labours of Hercules excel the same,
Much farther from true reasoning thou farest. 30
For what could hurt us now that mighty maw
Of Nemeaean Lion, or what the Boar
Who bristled in Arcadia? Or, again,
O what could Cretan Bull, or Hydra, pest
Of Lerna, fenced with vipers venomous? 35
Or what the triple-breasted power of her
The three-fold Geryon...
The sojourners in the Stymphalian fens
So dreadfully offend us, or the Steeds
Of Thracian Diomedes breathing fire 40
From out their nostrils off along the zones
Bistonian and Ismarian? And the Snake,
The dread fierce gazer, guardian of the golden
And gleaming apples of the Hesperides,
Coiled round the tree-trunk with tremendous bulk, 45
O what, again, could he inflict on us
Along the Atlantic shore and wastes of sea?–
Where neither one of us approacheth nigh
Nor no barbarian ventures. And the rest
Of all those monsters slain, even if alive, 50
Unconquered still, what injury could they do?
None, as I guess. For so the glutted earth
Swarms even now with savage beasts, even now
Is filled with anxious terrors through the woods
And mighty mountains and the forest deeps– 55
Quarters 'tis ours in general to avoid.
But lest the breast be purged, what conflicts then,
What perils, must bosom, in our own despite!
O then how great and keen the cares of lust
That split the man distraught! How great the fears! 60
And lo, the pride, grim greed, and wantonness–
How great the slaughters in their train! and lo,
Debaucheries and every breed of sloth!
Therefore that man who subjugated these,
And from the mind expelled, by words indeed, 65
Not arms, O shall it not be seemly him
To dignify by ranking with the gods?–

And all the more since he was wont to give,
Concerning the immortal gods themselves,
Many pronouncements with a tongue divine, 70
And to unfold by his pronouncements all
The nature of the world.

ARGUMENT OF THE BOOK AND NEW PROEM AGAINST A TELEOLOGICAL CONCEPT 75

And walking now
In his own footprints, I do follow through
His reasonings, and with pronouncements teach
The covenant whereby all things are framed, 80
How under that covenant they must abide
Nor ever prevail to abrogate the aeons'
Inexorable decrees,–how (as we've found),
In class of mortal objects, o'er all else,
The mind exists of earth-born frame create 85
And impotent unscathed to abide
Across the mighty aeons, and how come
In sleep those idol-apparitions,
That so befool intelligence when we
Do seem to view a man whom life has left. 90
Thus far we've gone; the order of my plan
Hath brought me now unto the point where I
Must make report how, too, the universe
Consists of mortal body, born in time,
And in what modes that congregated stuff 95
Established itself as earth and sky,
Ocean, and stars, and sun, and ball of moon;
And then what living creatures rose from out
The old telluric places, and what ones
Were never born at all; and in what mode 100
The human race began to name its things
And use the varied speech from man to man;
And in what modes hath bosomed in their breasts
That awe of gods, which halloweth in all lands
Fanes, altars, groves, lakes, idols of the gods. 105
Also I shall untangle by what power
The steersman nature guides the sun's courses,
And the meanderings of the moon, lest we,

Percase, should fancy that of own free will
They circle their perennial courses round, 110
Timing their motions for increase of crops
And living creatures, or lest we should think
They roll along by any plan of gods.
For even those men who have learned full well
That godheads lead a long life free of care, 115
If yet meanwhile they wonder by what plan
Things can go on (and chiefly yon high things
Observed o'erhead on the ethereal coasts),
Again are hurried back unto the fears
Of old religion and adopt again 120
Harsh masters, deemed almighty,–wretched men,
Unwitting what can be and what cannot,
And by what law to each its scope prescribed,
Its boundary stone that clings so deep in Time.

125

But for the rest,–lest we delay thee here
Longer by empty promises–behold,
Before all else, the seas, the lands, the sky:
O Memmius, their threefold nature, lo,
Their bodies three, three aspects so unlike, 130
Three frames so vast, a single day shall give
Unto annihilation! Then shall crash
That massive form and fabric of the world
Sustained so many aeons! Nor do I
Fail to perceive how strange and marvellous 135
This fact must strike the intellect of man,–
Annihilation of the sky and earth
That is to be,–and with what toil of words
'Tis mine to prove the same; as happens oft
When once ye offer to man's listening ears 140
Something before unheard of, but may not
Subject it to the view of eyes for him
Nor put it into hand–the sight and touch,
Whereby the opened highways of belief
Lead most directly into human breast 145
And regions of intelligence. But yet
I will speak out. The fact itself, perchance,
Will force belief in these my words, and thou
Mayst see, in little time, tremendously

With risen commotions of the lands all things 150
Quaking to pieces–which afar from us
May she, the steersman Nature, guide: and may
Reason, O rather than the fact itself,
Persuade us that all things can be o'erthrown
And sink with awful-sounding breakage down! 155

But ere on this I take a step to utter
Oracles holier and soundlier based
Than ever the Pythian pronounced for men
From out the tripod and the Delphian laurel, 160
I will unfold for thee with learned words
Many a consolation, lest perchance,
Still bridled by religion, thou suppose
Lands, sun, and sky, sea, constellations, moon,
Must dure forever, as of frame divine– 165
And so conclude that it is just that those,
(After the manner of the Giants), should all
Pay the huge penalties for monstrous crime,
Who by their reasonings do overshake
The ramparts of the universe and wish 170
There to put out the splendid sun of heaven,
Branding with mortal talk immortal things–
Though these same things are even so far removed
From any touch of deity and seem
So far unworthy of numbering with the gods, 175
That well they may be thought to furnish rather
A goodly instance of the sort of things
That lack the living motion, living sense.
For sure 'tis quite beside the mark to think
That judgment and the nature of the mind 180
In any kind of body can exist–
Just as in ether can't exist a tree,
Nor clouds in the salt sea, nor in the fields
Can fishes live, nor blood in timber be,
Nor sap in boulders: fixed and arranged 185
Where everything may grow and have its place.
Thus nature of mind cannot arise alone
Without the body, nor have its being far
From thews and blood. Yet if 'twere possible?–
Much rather might this very power of mind 190

Be in the head, the shoulders, or the heels,
And, born in any part soever, yet
In the same man, in the same vessel abide
But since within this body even of ours
Stands fixed and appears arranged sure 195
Where soul and mind can each exist and grow,
Deny we must the more that they can dure
Outside the body and the breathing form
In rotting clods of earth, in the sun's fire,
In water, or in ether's skiey coasts. 200
Therefore these things no whit are furnished
With sense divine, since never can they be
With life-force quickened.

Likewise, thou canst ne'er 205
Believe the sacred seats of gods are here
In any regions of this mundane world;
Indeed, the nature of the gods, so subtle,
So far removed from these our senses, scarce
Is seen even by intelligence of mind. 210
And since they've ever eluded touch and thrust
Of human hands, they cannot reach to grasp
Aught tangible to us. For what may not
Itself be touched in turn can never touch.
Wherefore, besides, also their seats must be 215
Unlike these seats of ours,–even subtle too,
As meet for subtle essence–as I'll prove
Hereafter unto thee with large discourse.
Further, to say that for the sake of men
They willed to prepare this world's magnificence, 220
And that 'tis therefore duty and behoof
To praise the work of gods as worthy praise,
And that 'tis sacrilege for men to shake
Ever by any force from out their seats
What hath been stablished by the Forethought old 225
To everlasting for races of mankind,
And that 'tis sacrilege to assault by words
And overtopple all from base to beam,–
Memmius, such notions to concoct and pile,
Is verily–to dote. Our gratefulness, 230
O what emoluments could it confer

Upon Immortals and upon the Blessed
That they should take a step to manage aught
For sake of us? Or what new factor could,
After so long a time, inveigle them– 235
The hitherto reposeful–to desire
To change their former life? For rather he
Whom old things chafe seems likely to rejoice
At new; but one that in fore-passed time
Hath chanced upon no ill, through goodly years, 240
O what could ever enkindle in such an one
Passion for strange experiment? Or what
The evil for us, if we had ne'er been born?–
As though, forsooth, in darkling realms and woe
Our life were lying till should dawn at last 245
The day-spring of creation! Whosoever
Hath been begotten wills perforce to stay
In life, so long as fond delight detains;
But whoso ne'er hath tasted love of life,
And ne'er was in the count of living things, 250
What hurts it him that he was never born?
Whence, further, first was planted in the gods
The archetype for gendering the world
And the fore-notion of what man is like,
So that they knew and pre-conceived with mind 255
Just what they wished to make? Or how were known
Ever the energies of primal germs,
And what those germs, by interchange of place,
Could thus produce, if nature's self had not
Given example for creating all? 260
For in such wise primordials of things,
Many in many modes, astir by blows
From immemorial aeons, in motion too
By their own weights, have evermore been wont
To be so borne along and in all modes 265
To meet together and to try all sorts
Which, by combining one with other, they
Are powerful to create, that thus it is
No marvel now, if they have also fallen
Into arrangements such, and if they've passed 270
Into vibrations such, as those whereby
This sum of things is carried on to-day

By fixed renewal. But knew I never what
The seeds primordial were, yet would I dare
This to affirm, even from deep judgments based 275
Upon the ways and conduct of the skies–
This to maintain by many a fact besides–
That in no wise the nature of all things
For us was fashioned by a power divine–
So great the faults it stands encumbered with. 280
First, mark all regions which are overarched
By the prodigious reaches of the sky:
One yawning part thereof the mountain-chains
And forests of the beasts do have and hold;
And cliffs, and desert fens, and wastes of sea 285
(Which sunder afar the beaches of the lands)
Possess it merely; and, again, thereof
Well-nigh two-thirds intolerable heat
And a perpetual fall of frost doth rob
From mortal kind. And what is left to till, 290
Even that the force of nature would o'errun
With brambles, did not human force oppose,–
Long wont for livelihood to groan and sweat
Over the two-pronged mattock and to cleave
The soil in twain by pressing on the plough. 295

Unless, by the ploughshare turning the fruitful clods
And kneading the mould, we quicken into birth,
[The crops] spontaneously could not come up
Into the free bright air. Even then sometimes, 300
When things acquired by the sternest toil
Are now in leaf, are now in blossom all,
Either the skiey sun with baneful heats
Parches, or sudden rains or chilling rime
Destroys, or flaws of winds with furious whirl 305
Torment and twist. Beside these matters, why
Doth nature feed and foster on land and sea
The dreadful breed of savage beasts, the foes
Of the human clan? Why do the seasons bring
Distempers with them? Wherefore stalks at large 310
Death, so untimely? Then, again, the babe,
Like to the castaway of the raging surf,
Lies naked on the ground, speechless, in want

Of every help for life, when nature first
Hath poured him forth upon the shores of light 315
With birth-pangs from within the mother's womb,
And with a plaintive wail he fills the place,–
As well befitting one for whom remains
In life a journey through so many ills.
But all the flocks and herds and all wild beasts 320
Come forth and grow, nor need the little rattles,
Nor must be treated to the humouring nurse's
Dear, broken chatter; nor seek they divers clothes
To suit the changing skies; nor need, in fine,
Nor arms, nor lofty ramparts, wherewithal 325
Their own to guard–because the earth herself
And nature, artificer of the world, bring forth
Aboundingly all things for all.

The World is not Eternal

And first,
Since body of earth and water, air's light breath,
And fiery exhalations (of which four
This sum of things is seen to be compact)
So all have birth and perishable frame, 5
Thus the whole nature of the world itself
Must be conceived as perishable too.
For, verily, those things of which we see
The parts and members to have birth in time
And perishable shapes, those same we mark 10
To be invariably born in time
And born to die. And therefore when I see
The mightiest members and the parts of this
Our world consumed and begot again,
'Tis mine to know that also sky above 15
And earth beneath began of old in time
And shall in time go under to disaster.

And lest in these affairs thou deemest me
To have seized upon this point by sleight to serve 20
My own caprice–because I have assumed
That earth and fire are mortal things indeed,
And have not doubted water and the air
Both perish too and have affirmed the same
To be again begotten and wax big– 25
Mark well the argument: in first place, lo,
Some certain parts of earth, grievously parched
By unremitting suns, and trampled on
By a vast throng of feet, exhale abroad

A powdery haze and flying clouds of dust, 30
Which the stout winds disperse in the whole air.
A part, moreover, of her sod and soil
Is summoned to inundation by the rains;
And rivers graze and gouge the banks away.
Besides, whatever takes a part its own 35
In fostering and increasing [aught]...

Is rendered back; and since, beyond a doubt,
Earth, the all-mother, is beheld to be
Likewise the common sepulchre of things, 40
Therefore thou seest her minished of her plenty,
And then again augmented with new growth.

And for the rest, that sea, and streams, and springs
Forever with new waters overflow, 45
And that perennially the fluids well,
Needeth no words–the mighty flux itself
Of multitudinous waters round about
Declareth this. But whatso water first
Streams up is ever straightway carried off, 50
And thus it comes to pass that all in all
There is no overflow; in part because
The burly winds (that over-sweep amain)
And skiey sun (that with his rays dissolves)
Do minish the level seas; in part because 55
The water is diffused underground
Through all the lands. The brine is filtered off,
And then the liquid stuff seeps back again
And all regathers at the river-heads,
Whence in fresh-water currents on it flows 60
Over the lands, adown the channels which
Were cleft erstwhile and erstwhile bore along
The liquid-footed floods.

Now, then, of air 65
I'll speak, which hour by hour in all its body
Is changed innumerably. For whatso'er
Streams up in dust or vapour off of things,
The same is all and always borne along
Into the mighty ocean of the air; 70

And did not air in turn restore to things
Bodies, and thus recruit them as they stream,
All things by this time had resolved been
And changed into air. Therefore it never
Ceases to be engendered off of things 75
And to return to things, since verily
In constant flux do all things stream.

Likewise,
The abounding well-spring of the liquid light, 80
The ethereal sun, doth flood the heaven o'er
With constant flux of radiance ever new,
And with fresh light supplies the place of light,
Upon the instant. For whatever effulgence
Hath first streamed off, no matter where it falls, 85
Is lost unto the sun. And this 'tis thine
To know from these examples: soon as clouds
Have first begun to under-pass the sun,
And, as it were, to rend the rays of light
In twain, at once the lower part of them 90
Is lost entire, and earth is overcast
Where'er the thunderheads are rolled along–
So know thou mayst that things forever need
A fresh replenishment of gleam and glow,
And each effulgence, foremost flashed forth, 95
Perisheth one by one. Nor otherwise
Can things be seen in sunlight, lest alway
The fountain-head of light supply new light.
Indeed your earthly beacons of the night,
The hanging lampions and the torches, bright 100
With darting gleams and dense with livid soot,
Do hurry in like manner to supply
With ministering heat new light amain;
Are all alive to quiver with their fires,–
Are so alive, that thus the light ne'er leaves 105
The spots it shines on, as if rent in twain:
So speedily is its destruction veiled
By the swift birth of flame from all the fires.
Thus, then, we must suppose that sun and moon
And stars dart forth their light from under-births 110
Ever and ever new, and whatso flames

First rise do perish always one by one–
Lest, haply, thou shouldst think they each endure
Inviolable.

115

Again, perceivest not
How stones are also conquered by Time?–
Not how the lofty towers ruin down,
And boulders crumble?–Not how shrines of gods
And idols crack outworn?–Nor how indeed 120
The holy Influence hath yet no power
There to postpone the Terminals of Fate,
Or headway make 'gainst Nature's fixed decrees?
Again, behold we not the monuments
Of heroes, now in ruins, asking us, 125
In their turn likewise, if we don't believe
They also age with eld? Behold we not
The rended basalt ruining amain
Down from the lofty mountains, powerless
To dure and dree the mighty forces there 130
Of finite time?–for they would never fall
Rended asudden, if from infinite Past
They had prevailed against all engin'ries
Of the assaulting aeons, with no crash.

135

Again, now look at This, which round, above,
Contains the whole earth in its one embrace:
If from itself it procreates all things–
As some men tell–and takes them to itself
When once destroyed, entirely must it be 140
Of mortal birth and body; for whate'er
From out itself giveth to other things
Increase and food, the same perforce must be
Minished, and then recruited when it takes
Things back into itself. 145

Besides all this,
If there had been no origin-in-birth
Of lands and sky, and they had ever been
The everlasting, why, ere Theban war 150
And obsequies of Troy, have other bards
Not also chanted other high affairs?

Whither have sunk so oft so many deeds
Of heroes? Why do those deeds live no more,
Ingrafted in eternal monuments 155
Of glory? Verily, I guess, because
The Sum is new, and of a recent date
The nature of our universe, and had
Not long ago its own exordium.
Wherefore, even now some arts are being still 160
Refined, still increased: now unto ships
Is being added many a new device;
And but the other day musician-folk
Gave birth to melic sounds of organing;
And, then, this nature, this account of things 165
Hath been discovered latterly, and I
Myself have been discovered only now,
As first among the first, able to turn
The same into ancestral Roman speech.
Yet if, percase, thou deemest that ere this 170
Existed all things even the same, but that
Perished the cycles of the human race
In fiery exhalations, or cities fell
By some tremendous quaking of the world,
Or rivers in fury, after constant rains, 175
Had plunged forth across the lands of earth
And whelmed the towns–then, all the more must thou
Confess, defeated by the argument,
That there shall be annihilation too
Of lands and sky. For at a time when things 180
Were being taxed by maladies so great,
And so great perils, if some cause more fell
Had then assailed them, far and wide they would
Have gone to disaster and supreme collapse.
And by no other reasoning are we 185
Seen to be mortal, save that all of us
Sicken in turn with those same maladies
With which have sickened in the past those men
Whom nature hath removed from life.

190

gain,
Whatever abides eternal must indeed
Either repel all strokes, because 'tis made

Of solid body, and permit no entrance
Of aught with power to sunder from within 195
The parts compact–as are those seeds of stuff
Whose nature we've exhibited before;
Or else be able to endure through time
For this: because they are from blows exempt,
As is the void, the which abides untouched, 200
Unsmit by any stroke; or else because
There is no room around, whereto things can,
As 'twere, depart in dissolution all,–
Even as the sum of sums eternal is,
Without or place beyond whereto things may 205
Asunder fly, or bodies which can smite,
And thus dissolve them by the blows of might.
But not of solid body, as I've shown,
Exists the nature of the world, because
In things is intermingled there a void; 210
Nor is the world yet as the void, nor are,
Moreover, bodies lacking which, percase,
Rising from out the infinite, can fell
With fury-whirlwinds all this sum of things,
Or bring upon them other cataclysm 215
Of peril strange; and yonder, too, abides
The infinite space and the profound abyss–
Whereinto, lo, the ramparts of the world
Can yet be shivered. Or some other power
Can pound upon them till they perish all. 220
Thus is the door of doom, O nowise barred
Against the sky, against the sun and earth
And deep-sea waters, but wide open stands
And gloats upon them, monstrous and agape.
Wherefore, again, 'tis needful to confess 225
That these same things are born in time; for things
Which are of mortal body could indeed
Never from infinite past until to-day
Have spurned the multitudinous assaults
Of the immeasurable aeons old. 230

Again, since battle so fiercely one with other
The four most mighty members the world,
Aroused in an all unholy war,

Seest not that there may be for them an end 235
Of the long strife?–Or when the skiey sun
And all the heat have won dominion o'er
The sucked-up waters all?–And this they try
Still to accomplish, though as yet they fail,–
For so aboundingly the streams supply 240
New store of waters that 'tis rather they
Who menace the world with inundations vast
From forth the unplumbed chasms of the sea.
But vain–since winds (that over-sweep amain)
And skiey sun (that with his rays dissolves) 245
Do minish the level seas and trust their power
To dry up all, before the waters can
Arrive at the end of their endeavouring.
Breathing such vasty warfare, they contend
In balanced strife the one with other still 250
Concerning mighty issues,–though indeed
The fire was once the more victorious,
And once–as goes the tale–the water won
A kingdom in the fields. For fire o'ermastered
And licked up many things and burnt away, 255
What time the impetuous horses of the Sun
Snatched Phaethon headlong from his skiey road
Down the whole ether and over all the lands.
But the omnipotent Father in keen wrath
Then with the sudden smite of thunderbolt 260
Did hurl the mighty-minded hero off
Those horses to the earth. And Sol, his sire,
Meeting him as he fell, caught up in hand
The ever-blazing lampion of the world,
And drave together the pell-mell horses there 265
And yoked them all a-tremble, and amain,
Steering them over along their own old road,
Restored the cosmos,–as forsooth we hear
From songs of ancient poets of the Greeks–
A tale too far away from truth, meseems. 270
For fire can win when from the infinite
Has risen a larger throng of particles
Of fiery stuff; and then its powers succumb,
Somehow subdued again, or else at last
It shrivels in torrid atmospheres the world. 275

And whilom water too began to win–
As goes the story–when it overwhelmed
The lives of men with billows; and thereafter,
When all that force of water-stuff which forth
From out the infinite had risen up 280
Did now retire, as somehow turned aside,
The rain-storms stopped, and streams their fury checked.

FORMATION OF THE WORLD AND ASTRONOMICAL QUESTIONS 285

But in what modes that conflux of first-stuff
Did found the multitudinous universe
Of earth, and sky, and the unfathomed deeps
Of ocean, and courses of the sun and moon, 290
I'll now in order tell. For of a truth
Neither by counsel did the primal germs
'Stablish themselves, as by keen act of mind,
Each in its proper place; nor did they make,
Forsooth, a compact how each germ should move; 295
But, lo, because primordials of things,
Many in many modes, astir by blows
From immemorial aeons, in motion too
By their own weights, have evermore been wont
To be so borne along and in all modes 300
To meet together and to try all sorts
Which, by combining one with other, they
Are powerful to create: because of this
It comes to pass that those primordials,
Diffused far and wide through mighty aeons, 305
The while they unions try, and motions too,
Of every kind, meet at the last amain,
And so become oft the commencements fit
Of mighty things–earth, sea, and sky, and race
Of living creatures. 310

In that long-ago
The wheel of the sun could nowhere be discerned
Flying far up with its abounding blaze,
Nor constellations of the mighty world, 315
Nor ocean, nor heaven, nor even earth nor air.

Nor aught of things like unto things of ours
Could then be seen–but only some strange storm
And a prodigious hurly-burly mass
Compounded of all kinds of primal germs, 320
Whose battling discords in disorder kept
Interstices, and paths, coherencies,
And weights, and blows, encounterings, and motions,
Because, by reason of their forms unlike
And varied shapes, they could not all thuswise 325
Remain conjoined nor harmoniously
Have interplay of movements. But from there
Portions began to fly asunder, and like
With like to join, and to block out a world,
And to divide its members and dispose 330
Its mightier parts–that is, to set secure
The lofty heavens from the lands, and cause
The sea to spread with waters separate,
And fires of ether separate and pure
Likewise to congregate apart. 335

For, lo,
First came together the earthy particles
(As being heavy and intertangled) there
In the mid-region, and all began to take 340
The lowest abodes; and ever the more they got
One with another intertangled, the more
They pressed from out their mass those particles
Which were to form the sea, the stars, the sun,
And moon, and ramparts of the mighty world– 345
For these consist of seeds more smooth and round
And of much smaller elements than earth.
And thus it was that ether, fraught with fire,
First broke away from out the earthen parts,
Athrough the innumerable pores of earth, 350
And raised itself aloft, and with itself
Bore lightly off the many starry fires;
And not far otherwise we often see

And the still lakes and the perennial streams 355
Exhale a mist, and even as earth herself
Is seen at times to smoke, when first at dawn

The light of the sun, the many-rayed, begins
To redden into gold, over the grass
Begemmed with dew. When all of these are brought 360
Together overhead, the clouds on high
With now concreted body weave a cover
Beneath the heavens. And thuswise ether too,
Light and diffusive, with concreted body
On all sides spread, on all sides bent itself 365
Into a dome, and, far and wide diffused
On unto every region on all sides,
Thus hedged all else within its greedy clasp.
Hard upon ether came the origins
Of sun and moon, whose globes revolve in air 370
Midway between the earth and mightiest ether,–
For neither took them, since they weighed too little
To sink and settle, but too much to glide
Along the upmost shores; and yet they are
In such a wise midway between the twain 375
As ever to whirl their living bodies round,
And ever to dure as parts of the wide Whole;
In the same fashion as certain members may
In us remain at rest, whilst others move.
When, then, these substances had been withdrawn, 380
Amain the earth, where now extend the vast
Cerulean zones of all the level seas,
Caved in, and down along the hollows poured
The whirlpools of her brine; and day by day
The more the tides of ether and rays of sun 385
On every side constrained into one mass
The earth by lashing it again, again,
Upon its outer edges (so that then,
Being thus beat upon, 'twas all condensed
About its proper centre), ever the more 290
The salty sweat, from out its body squeezed,
Augmented ocean and the fields of foam
By seeping through its frame, and all the more
Those many particles of heat and air
Escaping, began to fly aloft, and form, 295
By condensation there afar from earth,
The high refulgent circuits of the heavens.
The plains began to sink, and windy slopes

Of the high mountains to increase; for rocks
Could not subside, nor all the parts of ground 300
Settle alike to one same level there.

Thus, then, the massy weight of earth stood firm
With now concreted body, when (as 'twere)
All of the slime of the world, heavy and gross, 305
Had run together and settled at the bottom,
Like lees or bilge. Then ocean, then the air,
Then ether herself, the fraught-with-fire, were all
Left with their liquid bodies pure and free,
And each more lighter than the next below; 310
And ether, most light and liquid of the three,
Floats on above the long aerial winds,
Nor with the brawling of the winds of air
Mingles its liquid body. It doth leave
All there–those under-realms below her heights– 315
There to be overset in whirlwinds wild,–
Doth leave all there to brawl in wayward gusts,
Whilst, gliding with a fixed impulse still,
Itself it bears its fires along. For, lo,
That ether can flow thus steadily on, on, 320
With one unaltered urge, the Pontus proves–
That sea which floweth forth with fixed tides,
Keeping one onward tenor as it glides.

And that the earth may there abide at rest 325
In the mid-region of the world, it needs
Must vanish bit by bit in weight and lessen,
And have another substance underneath,
Conjoined to it from its earliest age
In linked unison with the vasty world's 330
Realms of the air in which it roots and lives.
On this account, the earth is not a load,
Nor presses down on winds of air beneath;
Even as unto a man his members be
Without all weight–the head is not a load 335
Unto the neck; nor do we feel the whole
Weight of the body to centre in the feet.
But whatso weights come on us from without,
Weights laid upon us, these harass and chafe,

Though often far lighter. For to such degree 340
It matters always what the innate powers
Of any given thing may be. The earth
Was, then, no alien substance fetched amain,
And from no alien firmament cast down
On alien air; but was conceived, like air, 345
In the first origin of this the world,
As a fixed portion of the same, as now
Our members are seen to be a part of us.

Besides, the earth, when of a sudden shook 350
By the big thunder, doth with her motion shake
All that's above her–which she ne'er could do
By any means, were earth not bounden fast
Unto the great world's realms of air and sky:
For they cohere together with common roots, 355
Conjoined both, even from their earliest age,
In linked unison. Aye, seest thou not
That this most subtle energy of soul
Supports our body, though so heavy a weight,–
Because, indeed, 'tis with it so conjoined 360
In linked unison? What power, in sum,
Can raise with agile leap our body aloft,
Save energy of mind which steers the limbs?
Now seest thou not how powerful may be
A subtle nature, when conjoined it is 365
With heavy body, as air is with the earth
Conjoined, and energy of mind with us?

Now let us sing what makes the stars to move.
In first place, if the mighty sphere of heaven 370
Revolveth round, then needs we must aver
That on the upper and the under pole
Presses a certain air, and from without
Confines them and encloseth at each end;
And that, moreover, another air above 375
Streams on athwart the top of the sphere and tends
In same direction as are rolled along
The glittering stars of the eternal world;
Or that another still streams on below
To whirl the sphere from under up and on 380

In opposite direction–as we see
The rivers turn the wheels and water-scoops.
It may be also that the heavens do all
Remain at rest, whilst yet are borne along
The lucid constellations; either because 385
Swift tides of ether are by sky enclosed,
And whirl around, seeking a passage out,
And everywhere make roll the starry fires
Through the Summanian regions of the sky;
Or else because some air, streaming along 390
From an eternal quarter off beyond,
Whileth the driven fires, or, then, because
The fires themselves have power to creep along,
Going wherever their food invites and calls,
And feeding their flaming bodies everywhere 395
Throughout the sky. Yet which of these is cause
In this our world 'tis hard to say for sure;
But what can be throughout the universe,
In divers worlds on divers plan create,
This only do I show, and follow on 400
To assign unto the motions of the stars
Even several causes which 'tis possible
Exist throughout the universal All;
Of which yet one must be the cause even here
Which maketh motion for our constellations. 405
Yet to decide which one of them it be
Is not the least the business of a man
Advancing step by cautious step, as I.

Nor can the sun's wheel larger be by much 410
Nor its own blaze much less than either seems
Unto our senses. For from whatso spaces
Fires have the power on us to cast their beams
And blow their scorching exhalations forth
Against our members, those same distances 415
Take nothing by those intervals away
From bulk of flames; and to the sight the fire
Is nothing shrunken. Therefore, since the heat
And the outpoured light of skiey sun
Arrive our senses and caress our limbs, 420
Form too and bigness of the sun must look

Even here from earth just as they really be,
So that thou canst scarce nothing take or add.
And whether the journeying moon illuminate
The regions round with bastard beams, or throw 425
From off her proper body her own light,–
Whichever it be, she journeys with a form
Naught larger than the form doth seem to be
Which we with eyes of ours perceive. For all
The far removed objects of our gaze 430
Seem through much air confused in their look
Ere minished in their bigness. Wherefore, moon,
Since she presents bright look and clear-cut form,
May there on high by us on earth be seen
Just as she is with extreme bounds defined, 435
And just of the size. And lastly, whatso fires
Of ether thou from earth beholdest, these
Thou mayst consider as possibly of size
The least bit less, or larger by a hair
Than they appear–since whatso fires we view 440
Here in the lands of earth are seen to change
From time to time their size to less or more
Only the least, when more or less away,
So long as still they bicker clear, and still
Their glow's perceived. 445

Nor need there be for men
Astonishment that yonder sun so small
Can yet send forth so great a light as fills
Oceans and all the lands and sky aflood, 450
And with its fiery exhalations steeps
The world at large. For it may be, indeed,
That one vast-flowing well-spring of the whole
Wide world from here hath opened and out-gushed,
And shot its light abroad; because thuswise 455
The elements of fiery exhalations
From all the world around together come,
And thuswise flow into a bulk so big
That from one single fountain-head may stream
This heat and light. And seest thou not, indeed, 460
How widely one small water-spring may wet
The meadow-lands at times and flood the fields?

'Tis even possible, besides, that heat
From forth the sun's own fire, albeit that fire
Be not a great, may permeate the air 465
With the fierce hot–if but, perchance, the air
Be of condition and so tempered then
As to be kindled, even when beat upon
Only by little particles of heat–
Just as we sometimes see the standing grain 470
Or stubble straw in conflagration all
From one lone spark. And possibly the sun,
Agleam on high with rosy lampion,
Possesses about him with invisible heats
A plenteous fire, by no effulgence marked, 475
So that he maketh, he, the Fraught-with-fire,
Increase to such degree the force of rays.

Nor is there one sure cause revealed to men
How the sun journeys from his summer haunts 480
On to the mid-most winter turning-points
In Capricorn, the thence reverting veers
Back to solstitial goals of Cancer; nor
How 'tis the moon is seen each month to cross
That very distance which in traversing 485
The sun consumes the measure of a year.
I say, no one clear reason hath been given
For these affairs. Yet chief in likelihood
Seemeth the doctrine which the holy thought
Of great Democritus lays down: that ever 490
The nearer the constellations be to earth
The less can they by whirling of the sky
Be borne along, because those skiey powers
Of speed aloft do vanish and decrease
In under-regions, and the sun is thus 495
Left by degrees behind amongst those signs
That follow after, since the sun he lies
Far down below the starry signs that blaze;
And the moon lags even tardier than the sun:
In just so far as is her course removed 500
From upper heaven and nigh unto the lands,
In just so far she fails to keep the pace
With starry signs above; for just so far

As feebler is the whirl that bears her on,
(Being, indeed, still lower than the sun), 505
In just so far do all the starry signs,
Circling around, o'ertake her and o'erpass.
Therefore it happens that the moon appears
More swiftly to return to any sign
Along the Zodiac, than doth the sun, 510
Because those signs do visit her again
More swiftly than they visit the great sun.
It can be also that two streams of air
Alternately at fixed periods
Blow out from transverse regions of the world, 515
Of which the one may thrust the sun away
From summer-signs to mid-most winter goals
And rigors of the cold, and the other then
May cast him back from icy shades of chill
Even to the heat-fraught regions and the signs 520
That blaze along the Zodiac. So, too,
We must suppose the moon and all the stars,
Which through the mighty and sidereal years
Roll round in mighty orbits, may be sped
By streams of air from regions alternate. 525
Seest thou not also how the clouds be sped
By contrary winds to regions contrary,
The lower clouds diversely from the upper?
Then, why may yonder stars in ether there
Along their mighty orbits not be borne 530
By currents opposite the one to other?

But night o'erwhelms the lands with vasty murk
Either when sun, after his diurnal course,
Hath walked the ultimate regions of the sky 535
And wearily hath panted forth his fires,
Shivered by their long journeying and wasted
By traversing the multitudinous air,
Or else because the self-same force that drave
His orb along above the lands compels 540
Him then to turn his course beneath the lands.
Matuta also at a fixed hour
Spreadeth the roseate morning out along
The coasts of heaven and deploys the light,

Either because the self-same sun, returning 545
Under the lands, aspires to seize the sky,
Striving to set it blazing with his rays
Ere he himself appear, or else because
Fires then will congregate and many seeds
Of heat are wont, even at a fixed time, 550
To stream together–gendering evermore
New suns and light. Just so the story goes
That from the Idaean mountain-tops are seen
Dispersed fires upon the break of day
Which thence combine, as 'twere, into one ball 555
And form an orb. Nor yet in these affairs
Is aught for wonder that these seeds of fire
Can thus together stream at time so fixed
And shape anew the splendour of the sun.
For many facts we see which come to pass 560
At fixed time in all things: burgeon shrubs
At fixed time, and at a fixed time
They cast their flowers; and Eld commands the teeth,
At time as surely fixed, to drop away,
And Youth commands the growing boy to bloom 565
With the soft down and let from both his cheeks
The soft beard fall. And lastly, thunder-bolts,
Snow, rains, clouds, winds, at seasons of the year
Nowise unfixed, all do come to pass.
For where, even from their old primordial start 570
Causes have ever worked in such a way,
And where, even from the world's first origin,
Thuswise have things befallen, so even now
After a fixed order they come round
In sequence also. 575

Likewise, days may wax
Whilst the nights wane, and daylight minished be
Whilst nights do take their augmentations,
Either because the self-same sun, coursing 580
Under the lands and over in two arcs,
A longer and a briefer, doth dispart
The coasts of ether and divides in twain
His orbit all unequally, and adds,
As round he's borne, unto the one half there 585

As much as from the other half he's ta'en,
Until he then arrives that sign of heaven
Where the year's node renders the shades of night
Equal unto the periods of light.
For when the sun is midway on his course 590
Between the blasts of northwind and of south,
Heaven keeps his two goals parted equally,
By virtue of the fixed position old
Of the whole starry Zodiac, through which
That sun, in winding onward, takes a year, 595
Illumining the sky and all the lands
With oblique light–as men declare to us
Who by their diagrams have charted well
Those regions of the sky which be adorned
With the arranged signs of Zodiac. 600
Or else, because in certain parts the air
Under the lands is denser, the tremulous
Bright beams of fire do waver tardily,
Nor easily can penetrate that air
Nor yet emerge unto their rising-place: 605
For this it is that nights in winter time
Do linger long, ere comes the many-rayed
Round Badge of the day. Or else because, as said,
In alternating seasons of the year
Fires, now more quick, and now more slow, are wont 610
To stream together,–the fires which make the sun
To rise in some one spot–therefore it is
That those men seem to speak the truth [who hold
A new sun is with each new daybreak born].
615

The moon she possibly doth shine because
Strook by the rays of sun, and day by day
May turn unto our gaze her light, the more
She doth recede from orb of sun, until,
Facing him opposite across the world, 620
She hath with full effulgence gleamed abroad,
And, at her rising as she soars above,
Hath there observed his setting; thence likewise
She needs must hide, as 'twere, her light behind
By slow degrees, the nearer now she glides, 625
Along the circle of the Zodiac,

From her far place toward fires of yonder sun,–
As those men hold who feign the moon to be
Just like a ball and to pursue a course
Betwixt the sun and earth. There is, again, 630
Some reason to suppose that moon may roll
With light her very own, and thus display
The varied shapes of her resplendence there.
For near her is, percase, another body,
Invisible, because devoid of light, 635
Borne on and gliding all along with her,
Which in three modes may block and blot her disk.
Again, she may revolve upon herself,
Like to a ball's sphere–if perchance that be–
One half of her dyed o'er with glowing light, 640
And by the revolution of that sphere
She may beget for us her varying shapes,
Until she turns that fiery part of her
Full to the sight and open eyes of men;
Thence by slow stages round and back she whirls, 645
Withdrawing thus the luminiferous part
Of her sphered mass and ball, as, verily,
The Babylonian doctrine of Chaldees,
Refuting the art of Greek astrologers,
Labours, in opposition, to prove sure– 650
As if, forsooth, the thing for which each fights,
Might not alike be true,–or aught there were
Wherefore thou mightest risk embracing one
More than the other notion. Then, again,
Why a new moon might not forevermore 655
Created be with fixed successions there
Of shapes and with configurations fixed,
And why each day that bright created moon
Might not miscarry and another be,
In its stead and place, engendered anew, 660
'Tis hard to show by reason, or by words
To prove absurd–since, lo, so many things
Can be create with fixed successions:
Spring-time and Venus come, and Venus' boy,
The winged harbinger, steps on before, 665
And hard on Zephyr's foot-prints Mother Flora,
Sprinkling the ways before them, filleth all

With colours and with odours excellent;
Whereafter follows arid Heat, and he
Companioned is by Ceres, dusty one, 670
And by the Etesian Breezes of the north;
Then cometh Autumn on, and with him steps
Lord Bacchus, and then other Seasons too
And other Winds do follow–the high roar
Of great Volturnus, and the Southwind strong 675
With thunder-bolts. At last earth's Shortest-Day
Bears on to men the snows and brings again
The numbing cold. And Winter follows her,
His teeth with chills a-chatter. Therefore, 'tis
The less a marvel, if at fixed time 680
A moon is thus begotten and again
At fixed time destroyed, since things so many
Can come to being thus at fixed time.
Likewise, the sun's eclipses and the moon's
Far occultations rightly thou mayst deem 685

As due to several causes. For, indeed,
Why should the moon be able to shut out
Earth from the light of sun, and on the side
To earthward thrust her high head under sun, 690
Opposing dark orb to his glowing beams–
And yet, at same time, one suppose the effect
Could not result from some one other body
Which glides devoid of light forevermore?
Again, why could not sun, in weakened state, 695
At fixed time for-lose his fires, and then,
When he has passed on along the air
Beyond the regions, hostile to his flames,
That quench and kill his fires, why could not he
Renew his light? And why should earth in turn 700
Have power to rob the moon of light, and there,
Herself on high, keep the sun hid beneath,
Whilst the moon glideth in her monthly course
Athrough the rigid shadows of the cone?–
And yet, at same time, some one other body 705
Not have the power to under-pass the moon,
Or glide along above the orb of sun,
Breaking his rays and outspread light asunder?

And still, if moon herself refulgent be
With her own sheen, why could she not at times 710
In some one quarter of the mighty world
Grow weak and weary, whilst she passeth through
Regions unfriendly to the beams her own?

Origins of Vegetable and Animal Life

And now to what remains!–Since I've resolved
By what arrangements all things come to pass
Through the blue regions of the mighty world,–
How we can know what energy and cause
Started the various courses of the sun 5
And the moon's goings, and by what far means
They can succumb, the while with thwarted light,
And veil with shade the unsuspecting lands,
When, as it were, they blink, and then again
With open eye survey all regions wide, 10
Resplendent with white radiance–I do now
Return unto the world's primeval age
And tell what first the soft young fields of earth
With earliest parturition had decreed
To raise in air unto the shores of light 15
And to entrust unto the wayward winds.
In the beginning, earth gave forth, around
The hills and over all the length of plains,
The race of grasses and the shining green;
The flowery meadows sparkled all aglow 20
With greening colour, and thereafter, lo,
Unto the divers kinds of trees was given
An emulous impulse mightily to shoot,
With a free rein, aloft into the air.
As feathers and hairs and bristles are begot 25
The first on members of the four-foot breeds
And on the bodies of the strong-y-winged,
Thus then the new Earth first of all put forth
Grasses and shrubs, and afterward begat

The mortal generations, there upsprung– 30
Innumerable in modes innumerable–
After diverging fashions. For from sky
These breathing-creatures never can have dropped,
Nor the land-dwellers ever have come up
Out of sea-pools of salt. How true remains, 35
How merited is that adopted name
Of earth–"The Mother!"–since from out the earth
Are all begotten. And even now arise
From out the loams how many living things–
Concreted by the rains and heat of the sun. 40
Wherefore 'tis less a marvel, if they sprang
In Long Ago more many, and more big,
Matured of those days in the fresh young years
Of earth and ether. First of all, the race
Of the winged ones and parti-coloured birds, 45
Hatched out in spring-time, left their eggs behind;
As now-a-days in summer tree-crickets
Do leave their shiny husks of own accord,
Seeking their food and living. Then it was
This earth of thine first gave unto the day 50
The mortal generations; for prevailed
Among the fields abounding hot and wet.
And hence, where any fitting spot was given,
There 'gan to grow womb-cavities, by roots
Affixed to earth. And when in ripened time 55
The age of the young within (that sought the air
And fled earth's damps) had burst these wombs, O then
Would Nature thither turn the pores of earth
And make her spurt from open veins a juice
Like unto milk; even as a woman now 60
Is filled, at child-bearing, with the sweet milk,
Because all that swift stream of aliment
Is thither turned unto the mother-breasts.
There earth would furnish to the children food;
Warmth was their swaddling cloth, the grass their bed 65
Abounding in soft down. Earth's newness then
Would rouse no dour spells of the bitter cold,
Nor extreme heats nor winds of mighty powers–
For all things grow and gather strength through time
In like proportions; and then earth was young. 70

Wherefore, again, again, how merited
Is that adopted name of Earth–The Mother!–
Since she herself begat the human race,
And at one well-nigh fixed time brought forth
Each breast that ranges raving round about 75
Upon the mighty mountains and all birds
Aerial with many a varied shape.
But, lo, because her bearing years must end,
She ceased, like to a woman worn by eld.
For lapsing aeons change the nature of 80
The whole wide world, and all things needs must take
One status after other, nor aught persists
Forever like itself. All things depart;
Nature she changeth all, compelleth all
To transformation. Lo, this moulders down, 85
A-slack with weary eld, and that, again,
Prospers in glory, issuing from contempt.
In suchwise, then, the lapsing aeons change
The nature of the whole wide world, and earth
Taketh one status after other. And what 90
She bore of old, she now can bear no longer,
And what she never bore, she can to-day.

In those days also the telluric world
Strove to beget the monsters that upsprung 95
With their astounding visages and limbs–
The Man-woman–a thing betwixt the twain,
Yet neither, and from either sex remote–
Some gruesome Boggles orphaned of the feet,
Some widowed of the hands, dumb Horrors too 100
Without a mouth, or blind Ones of no eye,
Or Bulks all shackled by their legs and arms
Cleaving unto the body fore and aft,
Thuswise, that never could they do or go,
Nor shun disaster, nor take the good they would. 105
And other prodigies and monsters earth
Was then begetting of this sort–in vain,
Since Nature banned with horror their increase,
And powerless were they to reach unto
The coveted flower of fair maturity, 110
Or to find aliment, or to intertwine

In works of Venus. For we see there must
Concur in life conditions manifold,
If life is ever by begetting life
To forge the generations one by one: 115
First, foods must be; and, next, a path whereby
The seeds of impregnation in the frame
May ooze, released from the members all;
Last, the possession of those instruments
Whereby the male with female can unite, 120
The one with other in mutual ravishments.

And in the ages after monsters died,
Perforce there perished many a stock, unable
By propagation to forge a progeny. 125
For whatsoever creatures thou beholdest
Breathing the breath of life, the same have been
Even from their earliest age preserved alive
By cunning, or by valour, or at least
By speed of foot or wing. And many a stock 130
Remaineth yet, because of use to man,
And so committed to man's guardianship.
Valour hath saved alive fierce lion-breeds
And many another terrorizing race,
Cunning the foxes, flight the antlered stags. 135
Light-sleeping dogs with faithful heart in breast,
However, and every kind begot from seed
Of beasts of draft, as, too, the woolly flocks
And horned cattle, all, my Memmius,
Have been committed to guardianship of men. 140
For anxiously they fled the savage beasts,
And peace they sought and their abundant foods,
Obtained with never labours of their own,
Which we secure to them as fit rewards
For their good service. But those beasts to whom 145
Nature has granted naught of these same things–
Beasts quite unfit by own free will to thrive
And vain for any service unto us
In thanks for which we should permit their kind
To feed and be in our protection safe– 150
Those, of a truth, were wont to be exposed,
Enshackled in the gruesome bonds of doom,

As prey and booty for the rest, until
Nature reduced that stock to utter death.

155

But Centaurs ne'er have been, nor can there be
Creatures of twofold stock and double frame,
Compact of members alien in kind,
Yet formed with equal function, equal force
In every bodily part–a fact thou mayst, 160
However dull thy wits, well learn from this:
The horse, when his three years have rolled away,
Flowers in his prime of vigour; but the boy
Not so, for oft even then he gropes in sleep
After the milky nipples of the breasts, 165
An infant still. And later, when at last
The lusty powers of horses and stout limbs,
Now weak through lapsing life, do fail with age,
Lo, only then doth youth with flowering years
Begin for boys, and clothe their ruddy cheeks 170
With the soft down. So never deem, percase,
That from a man and from the seed of horse,
The beast of draft, can Centaurs be composed
Or e'er exist alive, nor Scyllas be–
The half-fish bodies girdled with mad dogs– 175
Nor others of this sort, in whom we mark
Members discordant each with each; for ne'er
At one same time they reach their flower of age
Or gain and lose full vigour of their frame,
And never burn with one same lust of love, 180
And never in their habits they agree,
Nor find the same foods equally delightsome–
Sooth, as one oft may see the bearded goats
Batten upon the hemlock which to man
Is violent poison. Once again, since flame 185
Is wont to scorch and burn the tawny bulks
Of the great lions as much as other kinds
Of flesh and blood existing in the lands,
How could it be that she, Chimaera lone,
With triple body–fore, a lion she; 190
And aft, a dragon; and betwixt, a goat–
Might at the mouth from out the body belch
Infuriate flame? Wherefore, the man who feigns

Such beings could have been engendered
When earth was new and the young sky was fresh 195
(Basing his empty argument on new)
May babble with like reason many whims
Into our ears: he'll say, perhaps, that then
Rivers of gold through every landscape flowed,
That trees were wont with precious stones to flower, 200
Or that in those far aeons man was born
With such gigantic length and lift of limbs
As to be able, based upon his feet,
Deep oceans to bestride or with his hands
To whirl the firmament around his head. 205
For though in earth were many seeds of things
In the old time when this telluric world
First poured the breeds of animals abroad,
Still that is nothing of a sign that then
Such hybrid creatures could have been begot 210
And limbs of all beasts heterogeneous
Have been together knit; because, indeed,
The divers kinds of grasses and the grains
And the delightsome trees–which even now
Spring up abounding from within the earth– 215
Can still ne'er be begotten with their stems
Begrafted into one; but each sole thing
Proceeds according to its proper wont
And all conserve their own distinctions based
In nature's fixed decree. 220

Origins and Savage Period of Mankind

But mortal man
Was then far hardier in the old champaign,
As well he should be, since a hardier earth
Had him begotten; builded too was he
Of bigger and more solid bones within, 5
And knit with stalwart sinews through the flesh,
Nor easily seized by either heat or cold,
Or alien food or any ail or irk.
And whilst so many lustrums of the sun
Rolled on across the sky, men led a life 10
After the roving habit of wild beasts.
Not then were sturdy guiders of curved ploughs,
And none knew then to work the fields with iron,
Or plant young shoots in holes of delved loam,
Or lop with hooked knives from off high trees 15
The boughs of yester-year. What sun and rains
To them had given, what earth of own accord
Created then, was boon enough to glad
Their simple hearts. Mid acorn-laden oaks
Would they refresh their bodies for the nonce; 20
And the wild berries of the arbute-tree,
Which now thou seest to ripen purple-red
In winter time, the old telluric soil
Would bear then more abundant and more big.
And many coarse foods, too, in long ago 25
The blooming freshness of the rank young world
Produced, enough for those poor wretches there.

And rivers and springs would summon them of old
To slake the thirst, as now from the great hills
The water's down-rush calls aloud and far 30
The thirsty generations of the wild.
So, too, they sought the grottos of the Nymphs–
The woodland haunts discovered as they ranged–
From forth of which they knew that gliding rills
With gush and splash abounding laved the rocks, 35
The dripping rocks, and trickled from above
Over the verdant moss; and here and there
Welled up and burst across the open flats.
As yet they knew not to enkindle fire
Against the cold, nor hairy pelts to use 40
And clothe their bodies with the spoils of beasts;
But huddled in groves, and mountain-caves, and woods,
And 'mongst the thickets hid their squalid backs,
When driven to flee the lashings of the winds
And the big rains. Nor could they then regard 45
The general good, nor did they know to use
In common any customs, any laws:
Whatever of booty fortune unto each
Had proffered, each alone would bear away,
By instinct trained for self to thrive and live. 50
And Venus in the forests then would link
The lovers' bodies; for the woman yielded
Either from mutual flame, or from the man's
Impetuous fury and insatiate lust,
Or from a bribe–as acorn-nuts, choice pears, 55
Or the wild berries of the arbute-tree.
And trusting wondrous strength of hands and legs,
They'd chase the forest-wanderers, the beasts;
And many they'd conquer, but some few they fled,
A-skulk into their hiding-places... 60

With the flung stones and with the ponderous heft
Of gnarled branch. And by the time of night
O'ertaken, they would throw, like bristly boars,
Their wildman's limbs naked upon the earth, 65
Rolling themselves in leaves and fronded boughs.
Nor would they call with lamentations loud
Around the fields for daylight and the sun,

Quaking and wand'ring in shadows of the night;
But, silent and buried in a sleep, they'd wait 70
Until the sun with rosy flambeau brought
The glory to the sky. From childhood wont
Ever to see the dark and day begot
In times alternate, never might they be
Wildered by wild misgiving, lest a night 75
Eternal should possess the lands, with light
Of sun withdrawn forever. But their care
Was rather that the clans of savage beasts
Would often make their sleep-time horrible
For those poor wretches; and, from home y-driven, 80
They'd flee their rocky shelters at approach
Of boar, the spumy-lipped, or lion strong,
And in the midnight yield with terror up
To those fierce guests their beds of out-spread leaves.
85

And yet in those days not much more than now
Would generations of mortality
Leave the sweet light of fading life behind.
Indeed, in those days here and there a man,
More oftener snatched upon, and gulped by fangs, 90
Afforded the beasts a food that roared alive,
Echoing through groves and hills and forest-trees,
Even as he viewed his living flesh entombed
Within a living grave; whilst those whom flight
Had saved, with bone and body bitten, shrieked, 95
Pressing their quivering palms to loathsome sores,
With horrible voices for eternal death–
Until, forlorn of help, and witless what
Might medicine their wounds, the writhing pangs
Took them from life. But not in those far times 100
Would one lone day give over unto doom
A soldiery in thousands marching on
Beneath the battle-banners, nor would then
The ramping breakers of the main seas dash
Whole argosies and crews upon the rocks. 105
But ocean uprisen would often rave in vain,
Without all end or outcome, and give up
Its empty menacings as lightly too;
Nor soft seductions of a serene sea

Could lure by laughing billows any man 110
Out to disaster: for the science bold
Of ship-sailing lay dark in those far times.
Again, 'twas then that lack of food gave o'er
Men's fainting limbs to dissolution: now
'Tis plenty overwhelms. Unwary, they 115
Oft for themselves themselves would then outpour
The poison; now, with nicer art, themselves
They give the drafts to others.

Beginnings of Civilization

Afterwards,
When huts they had procured and pelts and fire,
And when the woman, joined unto the man,
Withdrew with him into one dwelling place,
5
Were known; and when they saw an offspring born
From out themselves, then first the human race
Began to soften. For 'twas now that fire
Rendered their shivering frames less staunch to bear,
Under the canopy of the sky, the cold; 10
And Love reduced their shaggy hardiness;
And children, with the prattle and the kiss,
Soon broke the parents' haughty temper down.
Then, too, did neighbours 'gin to league as friends,
Eager to wrong no more or suffer wrong, 15
And urged for children and the womankind
Mercy, of fathers, whilst with cries and gestures
They stammered hints how meet it was that all
Should have compassion on the weak. And still,
Though concord not in every wise could then 20
Begotten be, a good, a goodly part
Kept faith inviolate–or else mankind
Long since had been unutterably cut off,
And propagation never could have brought
The species down the ages. 25

Lest, perchance,
Concerning these affairs thou ponderest
In silent meditation, let me say

'Twas lightning brought primevally to earth 30
The fire for mortals, and from thence hath spread
O'er all the lands the flames of heat. For thus
Even now we see so many objects, touched
By the celestial flames, to flash aglow,
When thunderbolt has dowered them with heat. 35
Yet also when a many-branched tree,
Beaten by winds, writhes swaying to and fro,
Pressing 'gainst branches of a neighbour tree,
There by the power of mighty rub and rub
Is fire engendered; and at times out-flares 40
The scorching heat of flame, when boughs do chafe
Against the trunks. And of these causes, either
May well have given to mortal men the fire.
Next, food to cook and soften in the flame
The sun instructed, since so oft they saw 45
How objects mellowed, when subdued by warmth
And by the raining blows of fiery beams,
Through all the fields.

And more and more each day 50
Would men more strong in sense, more wise in heart,
Teach them to change their earlier mode and life
By fire and new devices. Kings began
Cities to found and citadels to set,
As strongholds and asylums for themselves, 55
And flocks and fields to portion for each man
After the beauty, strength, and sense of each–
For beauty then imported much, and strength
Had its own rights supreme. Thereafter, wealth
Discovered was, and gold was brought to light, 60
Which soon of honour stripped both strong and fair;
For men, however beautiful in form
Or valorous, will follow in the main
The rich man's party. Yet were man to steer
His life by sounder reasoning, he'd own 65
Abounding riches, if with mind content
He lived by thrift; for never, as I guess,
Is there a lack of little in the world.
But men wished glory for themselves and power
Even that their fortunes on foundations firm 70

Might rest forever, and that they themselves,
The opulent, might pass a quiet life–
In vain, in vain; since, in the strife to climb
On to the heights of honour, men do make
Their pathway terrible; and even when once 75
They reach them, envy like the thunderbolt
At times will smite, O hurling headlong down
To murkiest Tartarus, in scorn; for, lo,
All summits, all regions loftier than the rest,
Smoke, blasted as by envy's thunderbolts; 80
So better far in quiet to obey,
Than to desire chief mastery of affairs
And ownership of empires. Be it so;
And let the weary sweat their life-blood out
All to no end, battling in hate along 85
The narrow path of man's ambition;
Since all their wisdom is from others' lips,
And all they seek is known from what they've heard
And less from what they've thought. Nor is this folly
Greater to-day, nor greater soon to be, 90
Than' twas of old.

And therefore kings were slain,
And pristine majesty of golden thrones
And haughty sceptres lay o'erturned in dust; 95
And crowns, so splendid on the sovereign heads,
Soon bloody under the proletarian feet,
Groaned for their glories gone–for erst o'er-much
Dreaded, thereafter with more greedy zest
Trampled beneath the rabble heel. Thus things 100
Down to the vilest lees of brawling mobs
Succumbed, whilst each man sought unto himself
Dominion and supremacy. So next
Some wiser heads instructed men to found
The magisterial office, and did frame 105
Codes that they might consent to follow laws.
For humankind, o'er wearied with a life
Fostered by force, was ailing from its feuds;
And so the sooner of its own free will
Yielded to laws and strictest codes. For since 110
Each hand made ready in its wrath to take

A vengeance fiercer than by man's fair laws
Is now conceded, men on this account
Loathed the old life fostered by force. 'Tis thence
That fear of punishments defiles each prize 115
Of wicked days; for force and fraud ensnare
Each man around, and in the main recoil
On him from whence they sprung. Not easy 'tis
For one who violates by ugly deeds
The bonds of common peace to pass a life 120
Composed and tranquil. For albeit he 'scape
The race of gods and men, he yet must dread
'Twill not be hid forever–since, indeed,
So many, oft babbling on amid their dreams
Or raving in sickness, have betrayed themselves 125
(As stories tell) and published at last
Old secrets and the sins.

But nature 'twas
Urged men to utter various sounds of tongue 130
And need and use did mould the names of things,
About in same wise as the lack-speech years
Compel young children unto gesturings,
Making them point with finger here and there
At what's before them. For each creature feels 135
By instinct to what use to put his powers.
Ere yet the bull-calf's scarce begotten horns
Project above his brows, with them he 'gins
Enraged to butt and savagely to thrust.
But whelps of panthers and the lion's cubs 140
With claws and paws and bites are at the fray
Already, when their teeth and claws be scarce
As yet engendered. So again, we see
All breeds of winged creatures trust to wings
And from their fledgling pinions seek to get 145
A fluttering assistance. Thus, to think
That in those days some man apportioned round
To things their names, and that from him men learned
Their first nomenclature, is foolery.
For why could he mark everything by words 150
And utter the various sounds of tongue, what time
The rest may be supposed powerless

To do the same? And, if the rest had not
Already one with other used words,
Whence was implanted in the teacher, then, 155
Fore-knowledge of their use, and whence was given
To him alone primordial faculty
To know and see in mind what 'twas he willed?
Besides, one only man could scarce subdue
An overmastered multitude to choose 160
To get by heart his names of things. A task
Not easy 'tis in any wise to teach
And to persuade the deaf concerning what
'Tis needful for to do. For ne'er would they
Allow, nor ne'er in anywise endure 165
Perpetual vain dingdong in their ears
Of spoken sounds unheard before. And what,
At last, in this affair so wondrous is,
That human race (in whom a voice and tongue
Were now in vigour) should by divers words 170
Denote its objects, as each divers sense
Might prompt?–since even the speechless herds, aye, since
The very generations of wild beasts
Are wont dissimilar and divers sounds
To rouse from in them, when there's fear or pain, 175
And when they burst with joys. And this, forsooth,
'Tis thine to know from plainest facts: when first
Huge flabby jowls of mad Molossian hounds,
Baring their hard white teeth, begin to snarl,
They threaten, with infuriate lips peeled back, 180
In sounds far other than with which they bark
And fill with voices all the regions round.
And when with fondling tongue they start to lick
Their puppies, or do toss them round with paws,
Feigning with gentle bites to gape and snap, 185
They fawn with yelps of voice far other then
Than when, alone within the house, they bay,
Or whimpering slink with cringing sides from blows.
Again the neighing of the horse, is that
Not seen to differ likewise, when the stud 190
In buoyant flower of his young years raves,
Goaded by winged Love, amongst the mares,
And when with widening nostrils out he snorts

The call to battle, and when haply he
Whinnies at times with terror-quaking limbs? 195
Lastly, the flying race, the dappled birds,
Hawks, ospreys, sea-gulls, searching food and life
Amid the ocean billows in the brine,
Utter at other times far other cries
Than when they fight for food, or with their prey 200
Struggle and strain. And birds there are which change
With changing weather their own raucous songs–
As long-lived generations of the crows
Or flocks of rooks, when they be said to cry
For rain and water and to call at times 205
For winds and gales. Ergo, if divers moods
Compel the brutes, though speechless evermore,
To send forth divers sounds, O truly then
How much more likely 'twere that mortal men
In those days could with many a different sound 210
Denote each separate thing.

And now what cause
Hath spread divinities of gods abroad
Through mighty nations, and filled the cities full 215
Of the high altars, and led to practices
Of solemn rites in season–rites which still
Flourish in midst of great affairs of state
And midst great centres of man's civic life,
The rites whence still a poor mortality 220
Is grafted that quaking awe which rears aloft
Still the new temples of gods from land to land
And drives mankind to visit them in throngs
On holy days–'tis not so hard to give
Reason thereof in speech. Because, in sooth, 225
Even in those days would the race of man
Be seeing excelling visages of gods
With mind awake; and in his sleeps, yet more–
Bodies of wondrous growth. And, thus, to these
Would men attribute sense, because they seemed 230
To move their limbs and speak pronouncements high,
Befitting glorious visage and vast powers.
And men would give them an eternal life,
Because their visages forevermore

Were there before them, and their shapes remained, 235
And chiefly, however, because men would not think
Beings augmented with such mighty powers
Could well by any force o'ermastered be.
And men would think them in their happiness
Excelling far, because the fear of death 240
Vexed no one of them at all, and since
At same time in men's sleeps men saw them do
So many wonders, and yet feel therefrom
Themselves no weariness. Besides, men marked
How in a fixed order rolled around 245
The systems of the sky, and changed times
Of annual seasons, nor were able then
To know thereof the causes. Therefore 'twas
Men would take refuge in consigning all
Unto divinities, and in feigning all 250
Was guided by their nod. And in the sky
They set the seats and vaults of gods, because
Across the sky night and the moon are seen
To roll along–moon, day, and night, and night's
Old awesome constellations evermore, 255
And the night-wandering fireballs of the sky,
And flying flames, clouds, and the sun, the rains,
Snow and the winds, the lightnings, and the hail,
And the swift rumblings, and the hollow roar
Of mighty menacings forevermore. 260

O humankind unhappy!–when it ascribed
Unto divinities such awesome deeds,
And coupled thereto rigours of fierce wrath!
What groans did men on that sad day beget 265
Even for themselves, and O what wounds for us,
What tears for our children's children! Nor, O man,
Is thy true piety in this: with head
Under the veil, still to be seen to turn
Fronting a stone, and ever to approach 270
Unto all altars; nor so prone on earth
Forward to fall, to spread upturned palms
Before the shrines of gods, nor yet to dew
Altars with profuse blood of four-foot beasts,
Nor vows with vows to link. But rather this: 275

To look on all things with a master eye
And mind at peace. For when we gaze aloft
Upon the skiey vaults of yon great world
And ether, fixed high o'er twinkling stars,
And into our thought there come the journeyings 280
Of sun and moon, O then into our breasts,
O'erburdened already with their other ills,
Begins forthwith to rear its sudden head
One more misgiving: lest o'er us, percase,
It be the gods' immeasurable power 285
That rolls, with varied motion, round and round
The far white constellations. For the lack
Of aught of reasons tries the puzzled mind:
Whether was ever a birth-time of the world,
And whether, likewise, any end shall be 290
How far the ramparts of the world can still
Outstand this strain of ever-roused motion,
Or whether, divinely with eternal weal
Endowed, they can through endless tracts of age
Glide on, defying the o'er-mighty powers 295
Of the immeasurable ages. Lo,
What man is there whose mind with dread of gods
Cringes not close, whose limbs with terror-spell
Crouch not together, when the parched earth
Quakes with the horrible thunderbolt amain, 300
And across the mighty sky the rumblings run?
Do not the peoples and the nations shake,
And haughty kings do they not hug their limbs,
Strook through with fear of the divinities,
Lest for aught foully done or madly said 305
The heavy time be now at hand to pay?
When, too, fierce force of fury-winds at sea
Sweepeth a navy's admiral down the main
With his stout legions and his elephants,
Doth he not seek the peace of gods with vows, 310
And beg in prayer, a-tremble, lulled winds
And friendly gales?–in vain, since, often up-caught
In fury-cyclones, is he borne along,
For all his mouthings, to the shoals of doom.
Ah, so irrevocably some hidden power 315
Betramples forevermore affairs of men,

And visibly grindeth with its heel in mire
The lictors' glorious rods and axes dire,
Having them in derision! Again, when earth
From end to end is rocking under foot, 320
And shaken cities ruin down, or threaten
Upon the verge, what wonder is it then
That mortal generations abase themselves,
And unto gods in all affairs of earth
Assign as last resort almighty powers 325
And wondrous energies to govern all?

Now for the rest: copper and gold and iron
Discovered were, and with them silver's weight
And power of lead, when with prodigious heat 330
The conflagrations burned the forest trees
Among the mighty mountains, by a bolt
Of lightning from the sky, or else because
Men, warring in the woodlands, on their foes
Had hurled fire to frighten and dismay, 335
Or yet because, by goodness of the soil
Invited, men desired to clear rich fields
And turn the countryside to pasture-lands,
Or slay the wild and thrive upon the spoils.
(For hunting by pit-fall and by fire arose 340
Before the art of hedging the covert round
With net or stirring it with dogs of chase.)
Howso the fact, and from what cause soever
The flamy heat with awful crack and roar
Had there devoured to their deepest roots 345
The forest trees and baked the earth with fire,
Then from the boiling veins began to ooze
O rivulets of silver and of gold,
Of lead and copper too, collecting soon
Into the hollow places of the ground. 350
And when men saw the cooled lumps anon
To shine with splendour-sheen upon the ground,
Much taken with that lustrous smooth delight,
They 'gan to pry them out, and saw how each
Had got a shape like to its earthy mould. 355
Then would it enter their heads how these same lumps,
If melted by heat, could into any form

Or figure of things be run, and how, again,
If hammered out, they could be nicely drawn
To sharpest points or finest edge, and thus 360
Yield to the forgers tools and give them power
To chop the forest down, to hew the logs,
To shave the beams and planks, besides to bore
And punch and drill. And men began such work
At first as much with tools of silver and gold 365
As with the impetuous strength of the stout copper;
But vainly–since their over-mastered power
Would soon give way, unable to endure,
Like copper, such hard labour. In those days
Copper it was that was the thing of price; 370
And gold lay useless, blunted with dull edge.
Now lies the copper low, and gold hath come
Unto the loftiest honours. Thus it is
That rolling ages change the times of things:
What erst was of a price, becomes at last 375
A discard of no honour; whilst another
Succeeds to glory, issuing from contempt,
And day by day is sought for more and more,
And, when 'tis found, doth flower in men's praise,
Objects of wondrous honour. 380

Now, Memmius,
How nature of iron discovered was, thou mayst
Of thine own self divine. Man's ancient arms
Were hands, and nails and teeth, stones too and boughs– 385
Breakage of forest trees–and flame and fire,
As soon as known. Thereafter force of iron
And copper discovered was; and copper's use
Was known ere iron's, since more tractable
Its nature is and its abundance more. 390
With copper men to work the soil began,
With copper to rouse the hurly waves of war,
To straw the monstrous wounds, and seize away
Another's flocks and fields. For unto them,
Thus armed, all things naked of defence 395
Readily yielded. Then by slow degrees
The sword of iron succeeded, and the shape
Of brazen sickle into scorn was turned:

With iron to cleave the soil of earth they 'gan,
And the contentions of uncertain war 400
Were rendered equal.

And, lo, man was wont
Armed to mount upon the ribs of horse
And guide him with the rein, and play about 405
With right hand free, oft times before he tried
Perils of war in yoked chariot;
And yoked pairs abreast came earlier
Than yokes of four, or scythed chariots
Whereinto clomb the men-at-arms. And next 410
The Punic folk did train the elephants–
Those curst Lucanian oxen, hideous,
The serpent-handed, with turrets on their bulks–
To dure the wounds of war and panic-strike
The mighty troops of Mars. Thus Discord sad 415
Begat the one Thing after other, to be
The terror of the nations under arms,
And day by day to horrors of old war
She added an increase.

420

Bulls, too, they tried
In war's grim business; and essayed to send
Outrageous boars against the foes. And some
Sent on before their ranks puissant lions
With armed trainers and with masters fierce 425
To guide and hold in chains–and yet in vain,
Since fleshed with pell-mell slaughter, fierce they flew,
And blindly through the squadrons havoc wrought,
Shaking the frightful crests upon their heads,
Now here, now there. Nor could the horsemen calm 430
Their horses, panic-breasted at the roar,
And rein them round to front the foe. With spring
The infuriate she-lions would up-leap
Now here, now there; and whoso came apace
Against them, these they'd rend across the face; 435
And others unwitting from behind they'd tear
Down from their mounts, and twining round them, bring
Tumbling to earth, o'ermastered by the wound,
And with those powerful fangs and hooked claws

Fasten upon them. Bulls would toss their friends, 440
And trample under foot, and from beneath
Rip flanks and bellies of horses with their horns,
And with a threat'ning forehead jam the sod;
And boars would gore with stout tusks their allies,
Splashing in fury their own blood on spears 445
Splintered in their own bodies, and would fell
In rout and ruin infantry and horse.
For there the beasts-of-saddle tried to scape
The savage thrusts of tusk by shying off,
Or rearing up with hoofs a-paw in air. 450
In vain–since there thou mightest see them sink,
Their sinews severed, and with heavy fall
Bestrew the ground. And such of these as men
Supposed well-trained long ago at home,
Were in the thick of action seen to foam 455
In fury, from the wounds, the shrieks, the flight,
The panic, and the tumult; nor could men
Aught of their numbers rally. For each breed
And various of the wild beasts fled apart
Hither or thither, as often in wars to-day 460
Flee those Lucanian oxen, by the steel
Grievously mangled, after they have wrought
Upon their friends so many a dreadful doom.
(If 'twas, indeed, that thus they did at all:
But scarcely I'll believe that men could not 465
With mind foreknow and see, as sure to come,
Such foul and general disaster.–This
We, then, may hold as true in the great All,
In divers worlds on divers plan create,–
Somewhere afar more likely than upon 470
One certain earth.) But men chose this to do
Less in the hope of conquering than to give
Their enemies a goodly cause of woe,
Even though thereby they perished themselves,
Since weak in numbers and since wanting arms. 475

Now, clothes of roughly inter-plaited strands
Were earlier than loom-wove coverings;
The loom-wove later than man's iron is,
Since iron is needful in the weaving art, 480

Nor by no other means can there be wrought
Such polished tools–the treadles, spindles, shuttles,
And sounding yarn-beams. And nature forced the men,
Before the woman kind, to work the wool:
For all the male kind far excels in skill, 485
And cleverer is by much–until at last
The rugged farmer folk jeered at such tasks,
And so were eager soon to give them o'er
To women's hands, and in more hardy toil
To harden arms and hands. 490

But nature herself,
Mother of things, was the first seed-sower
And primal grafter; since the berries and acorns,
Dropping from off the trees, would there beneath 495
Put forth in season swarms of little shoots;
Hence too men's fondness for ingrafting slips
Upon the boughs and setting out in holes
The young shrubs o'er the fields. Then would they try
Ever new modes of tilling their loved crofts, 500
And mark they would how earth improved the taste
Of the wild fruits by fond and fostering care.
And day by day they'd force the woods to move
Still higher up the mountain, and to yield
The place below for tilth, that there they might, 505
On plains and uplands, have their meadow-plats,
Cisterns and runnels, crops of standing grain,
And happy vineyards, and that all along
O'er hillocks, intervales, and plains might run
The silvery-green belt of olive-trees, 510
Marking the plotted landscape; even as now
Thou seest so marked with varied loveliness
All the terrain which men adorn and plant
With rows of goodly fruit-trees and hedge round
With thriving shrubberies sown. 515

But by the mouth
To imitate the liquid notes of birds
Was earlier far 'mongst men than power to make,
By measured song, melodious verse and give 520
Delight to ears. And whistlings of the wind

Athrough the hollows of the reeds first taught
The peasantry to blow into the stalks
Of hollow hemlock-herb. Then bit by bit
They learned sweet plainings, such as pipe out-pours, 525
Beaten by finger-tips of singing men,
When heard through unpathed groves and forest deeps
And woodsy meadows, through the untrod haunts
Of shepherd folk and spots divinely still.
Thus time draws forward each and everything 530
Little by little unto the midst of men,
And reason uplifts it to the shores of light.
These tunes would soothe and glad the minds of mortals
When sated with food,–for songs are welcome then.
And often, lounging with friends in the soft grass 535
Beside a river of water, underneath
A big tree's branches, merrily they'd refresh
Their frames, with no vast outlay–most of all
If the weather were smiling and the times of the year
Were painting the green of the grass around with flowers. 540
Then jokes, then talk, then peals of jollity
Would circle round; for then the rustic muse
Was in her glory; then would antic Mirth
Prompt them to garland head and shoulders about
With chaplets of intertwined flowers and leaves, 545
And to dance onward, out of tune, with limbs
Clownishly swaying, and with clownish foot
To beat our mother earth–from whence arose
Laughter and peals of jollity, for, lo,
Such frolic acts were in their glory then, 550
Being more new and strange. And wakeful men
Found solaces for their unsleeping hours
In drawing forth variety of notes,
In modulating melodies, in running
With puckered lips along the tuned reeds, 555
Whence, even in our day do the watchmen guard
These old traditions, and have learned well
To keep true measure. And yet they no whit
Do get a larger fruit of gladsomeness
Than got the woodland aborigines 560
In olden times. For what we have at hand–
If theretofore naught sweeter we have known–

That chiefly pleases and seems best of all;
But then some later, likely better, find
Destroys its worth and changes our desires 565
Regarding good of yesterday.

And thus
Began the loathing of the acorn; thus
Abandoned were those beds with grasses strewn 570
And with the leaves beladen. Thus, again,
Fell into new contempt the pelts of beasts–
Erstwhile a robe of honour, which, I guess,
Aroused in those days envy so malign
That the first wearer went to woeful death 575
By ambuscades,–and yet that hairy prize,
Rent into rags by greedy foemen there
And splashed by blood, was ruined utterly
Beyond all use or vantage. Thus of old
'Twas pelts, and of to-day 'tis purple and gold 580
That cark men's lives with cares and weary with war.
Wherefore, methinks, resides the greater blame
With us vain men to-day: for cold would rack,
Without their pelts, the naked sons of earth;
But us it nothing hurts to do without 585
The purple vestment, broidered with gold
And with imposing figures, if we still
Make shift with some mean garment of the Plebs.
So man in vain futilities toils on
Forever and wastes in idle cares his years– 590
Because, of very truth, he hath not learnt
What the true end of getting is, nor yet
At all how far true pleasure may increase.
And 'tis desire for better and for more
Hath carried by degrees mortality 595
Out onward to the deep, and roused up
From the far bottom mighty waves of war.

But sun and moon, those watchmen of the world,
With their own lanterns traversing around 600
The mighty, the revolving vault, have taught
Unto mankind that seasons of the years
Return again, and that the Thing takes place

After a fixed plan and order fixed.
605
Already would they pass their life, hedged round
By the strong towers; and cultivate an earth
All portioned out and boundaried; already
Would the sea flower and sail-winged ships;
Already men had, under treaty pacts, 610
Confederates and allies, when poets began
To hand heroic actions down in verse;
Nor long ere this had letters been devised–
Hence is our age unable to look back
On what has gone before, except where reason 615
Shows us a footprint.

Sailings on the seas,
Tillings of fields, walls, laws, and arms, and roads,
Dress and the like, all prizes, all delights 620
Of finer life, poems, pictures, chiselled shapes
Of polished sculptures–all these arts were learned
By practice and the mind's experience,
As men walked forward step by eager step.
Thus time draws forward each and everything 625
Little by little into the midst of men,
And reason uplifts it to the shores of light.
For one thing after other did men see
Grow clear by intellect, till with their arts
They've now achieved the supreme pinnacle. 630

Book Six

Proem

'Twas Athens first, the glorious in name,
That whilom gave to hapless sons of men
The sheaves of harvest, and re-ordered life,
And decreed laws; and she the first that gave
Life its sweet solaces, when she begat 5
A man of heart so wise, who whilom poured
All wisdom forth from his truth-speaking mouth;
The glory of whom, though dead, is yet to-day,
Because of those discoveries divine
Renowned of old, exalted to the sky. 10
For when saw he that well-nigh everything
Which needs of man most urgently require
Was ready to hand for mortals, and that life,
As far as might be, was established safe,
That men were lords in riches, honour, praise, 15
And eminent in goodly fame of sons,
And that they yet, O yet, within the home,
Still had the anxious heart which vexed life
Unpausingly with torments of the mind,
And raved perforce with angry plaints, then he, 20
Then he, the master, did perceive that 'twas
The vessel itself which worked the bane, and all,
However wholesome, which from here or there
Was gathered into it, was by that bane
Spoilt from within,–in part, because he saw 25
The vessel so cracked and leaky that nowise

'T could ever be filled to brim; in part because
He marked how it polluted with foul taste
Whate'er it got within itself. So he,
The master, then by his truth-speaking words, 30
Purged the breasts of men, and set the bounds
Of lust and terror, and exhibited
The supreme good whither we all endeavour,
And showed the path whereby we might arrive
Thereunto by a little cross-cut straight, 35
And what of ills in all affairs of mortals
Upsprang and flitted deviously about
(Whether by chance or force), since nature thus
Had destined; and from out what gates a man
Should sally to each combat. And he proved 40
That mostly vainly doth the human race
Roll in its bosom the grim waves of care.
For just as children tremble and fear all
In the viewless dark, so even we at times
Dread in the light so many things that be 45
No whit more fearsome than what children feign,
Shuddering, will be upon them in the dark.
This terror then, this darkness of the mind,
Not sunrise with its flaring spokes of light,
Nor glittering arrows of morning can disperse, 50
But only nature's aspect and her law.
Wherefore the more will I go on to weave
In verses this my undertaken task.

And since I've taught thee that the world's great vaults 55
Are mortal and that sky is fashioned
Of frame e'en born in time, and whatsoe'er
Therein go on and must perforce go on

The most I have unravelled; what remains 60
Do thou take in, besides; since once for all
To climb into that chariot' renowned

Of winds arise; and they appeased are
So that all things again... 65

Which were, are changed now, with fury stilled;
All other movements through the earth and sky
Which mortals gaze upon (O anxious oft
In quaking thoughts!), and which abase their minds 70
With dread of deities and press them crushed
Down to the earth, because their ignorance
Of cosmic causes forces them to yield
All things unto the empery of gods
And to concede the kingly rule to them. 75
For even those men who have learned full well
That godheads lead a long life free of care,
If yet meanwhile they wonder by what plan
Things can go on (and chiefly yon high things
Observed o'erhead on the ethereal coasts), 80
Again are hurried back unto the fears
Of old religion and adopt again
Harsh masters, deemed almighty,–wretched men,
Unwitting what can be and what cannot,
And by what law to each its scope prescribed, 85
Its boundary stone that clings so deep in Time.
Wherefore the more are they borne wandering on
By blindfold reason. And, Memmius, unless
From out thy mind thou spuest all of this
And casteth far from thee all thoughts which be 90
Unworthy gods and alien to their peace,
Then often will the holy majesties
Of the high gods be harmful unto thee,
As by thy thought degraded,–not, indeed,
That essence supreme of gods could be by this 95
So outraged as in wrath to thirst to seek
Revenges keen; but even because thyself
Thou plaguest with the notion that the gods,
Even they, the Calm Ones in serene repose,
Do roll the mighty waves of wrath on wrath; 100
Nor wilt thou enter with a serene breast
Shrines of the gods; nor wilt thou able be
In tranquil peace of mind to take and know
Those images which from their holy bodies
Are carried into intellects of men, 105
As the announcers of their form divine.
What sort of life will follow after this

'Tis thine to see. But that afar from us
Veriest reason may drive such life away,
Much yet remains to be embellished yet 110
In polished verses, albeit hath issued forth
So much from me already; lo, there is
The law and aspect of the sky to be
By reason grasped; there are the tempest times
And the bright lightnings to be hymned now– 115
Even what they do and from what cause soe'er
They're borne along–that thou mayst tremble not,
Marking off regions of prophetic skies
For auguries, O foolishly distraught
Even as to whence the flying flame hath come, 120
Or to which half of heaven it turns, or how
Through walled places it hath wound its way,
Or, after proving its dominion there,
How it hath speeded forth from thence amain–
Whereof nowise the causes do men know, 125
And think divinities are working there.
Do thou, Calliope, ingenious Muse,
Solace of mortals and delight of gods,
Point out the course before me, as I race
On to the white line of the utmost goal, 130
That I may get with signal praise the crown,
With thee my guide!

Great Meteorological Phenomena, Etc.

And so in first place, then,
With thunder are shaken the blue deeps of heaven,
Because the ethereal clouds, scudding aloft,
Together clash, what time 'gainst one another
The winds are battling. For never a sound there comes 5
From out the serene regions of the sky;
But wheresoever in a host more dense
The clouds foregather, thence more often comes
A crash with mighty rumbling. And, again,
Clouds cannot be of so condensed a frame 10
As stones and timbers, nor again so fine
As mists and flying smoke; for then perforce
They'd either fall, borne down by their brute weight,
Like stones, or, like the smoke, they'd powerless be
To keep their mass, or to retain within 15
Frore snows and storms of hail. And they give forth
O'er skiey levels of the spreading world
A sound on high, as linen-awning, stretched
O'er mighty theatres, gives forth at times
A cracking roar, when much 'tis beaten about 20
Betwixt the poles and cross-beams. Sometimes, too,
Asunder rent by wanton gusts, it raves
And imitates the tearing sound of sheets
Of paper–even this kind of noise thou mayst
In thunder hear–or sound as when winds whirl 25
With lashings and do buffet about in air
A hanging cloth and flying paper-sheets.

For sometimes, too, it chances that the clouds
Cannot together crash head-on, but rather
Move side-wise and with motions contrary 30
Graze each the other's body without speed,
From whence that dry sound grateth on our ears,
So long drawn-out, until the clouds have passed
From out their close positions.

35

And, again,
In following wise all things seem oft to quake
At shock of heavy thunder, and mightiest walls
Of the wide reaches of the upper world
There on the instant to have sprung apart, 40
Riven asunder, what time a gathered blast
Of the fierce hurricane hath all at once
Twisted its way into a mass of clouds,
And, there enclosed, ever more and more
Compelleth by its spinning whirl the cloud 45
To grow all hollow with a thickened crust
Surrounding; for thereafter, when the force
And the keen onset of the wind have weakened
That crust, lo, then the cloud, to-split in twain,
Gives forth a hideous crash with bang and boom. 50
No marvel this; since oft a bladder small,
Filled up with air, will, when of sudden burst,
Give forth a like large sound.

There's reason, too, 55
Why clouds make sounds, as through them blow the winds:
We see, borne down the sky, oft shapes of clouds
Rough-edged or branched many forky ways;
And 'tis the same, as when the sudden flaws
Of north-west wind through the dense forest blow, 60
Making the leaves to sough and limbs to crash.
It happens too at times that roused force
Of the fierce hurricane to-rends the cloud,
Breaking right through it by a front assault;
For what a blast of wind may do up there 65
Is manifest from facts when here on earth
A blast more gentle yet uptwists tall trees
And sucks them madly from their deepest roots.

Besides, among the clouds are waves, and these
Give, as they roughly break, a rumbling roar; 70
As when along deep streams or the great sea
Breaks the loud surf. It happens, too, whenever
Out from one cloud into another falls
The fiery energy of thunderbolt,
That straightaway the cloud, if full of wet, 75
Extinguishes the fire with mighty noise;
As iron, white from the hot furnaces,
Sizzles, when speedily we've plunged its glow
Down the cold water. Further, if a cloud
More dry receive the fire, 'twill suddenly 80
Kindle to flame and burn with monstrous sound,
As if a flame with whirl of winds should range
Along the laurel-tressed mountains far,
Upburning with its vast assault those trees;
Nor is there aught that in the crackling flame 85
Consumes with sound more terrible to man
Than Delphic laurel of Apollo lord.
Oft, too, the multitudinous crash of ice
And down-pour of swift hail gives forth a sound
Among the mighty clouds on high; for when 90
The wind hath packed them close, each mountain mass
Of rain-cloud, there congealed utterly
And mixed with hail-stones, breaks and booms...

Likewise, it lightens, when the clouds have struck, 95
By their collision, forth the seeds of fire:
As if a stone should smite a stone or steel,
For light then too leaps forth and fire then scatters
The shining sparks. But with our ears we get
The thunder after eyes behold the flash, 100
Because forever things arrive the ears
More tardily than the eyes–as thou mayst see
From this example too: when markest thou
Some man far yonder felling a great tree
With double-edged ax, it comes to pass 105
Thine eye beholds the swinging stroke before
The blow gives forth a sound athrough thine ears:
Thus also we behold the flashing ere
We hear the thunder, which discharged is

At same time with the fire and by same cause, 110
Born of the same collision.

In following wise
The clouds suffuse with leaping light the lands,
And the storm flashes with tremulous elan: 115
When the wind hath invaded a cloud, and, whirling there,
Hath wrought (as I have shown above) the cloud
Into a hollow with a thickened crust,
It becomes hot of own velocity:
Just as thou seest how motion will o'erheat 120
And set ablaze all objects,–verily
A leaden ball, hurtling through length of space,
Even melts. Therefore, when this same wind a-fire
Hath split black cloud, it scatters the fire-seeds,
Which, so to say, have been pressed out by force 125
Of sudden from the cloud;–and these do make
The pulsing flashes of flame; thence followeth
The detonation which attacks our ears
More tardily than aught which comes along
Unto the sight of eyeballs. This takes place– 130
As know thou mayst–at times when clouds are dense
And one upon the other piled aloft
With wonderful upheavings–nor be thou
Deceived because we see how broad their base
From underneath, and not how high they tower. 135
For make thine observations at a time
When winds shall bear athwart the horizon's blue
Clouds like to mountain-ranges moving on,
Or when about the sides of mighty peaks
Thou seest them one upon the other massed 140
And burdening downward, anchored in high repose,
With the winds sepulchred on all sides round:
Then canst thou know their mighty masses, then
Canst view their caverns, as if builded there
Of beetling crags; which, when the hurricanes 145
In gathered storm have filled utterly,
Then, prisoned in clouds, they rave around
With mighty roarings, and within those dens
Bluster like savage beasts, and now from here,
And now from there, send growlings through the clouds, 150

And seeking an outlet, whirl themselves about,
And roll from 'mid the clouds the seeds of fire,
And heap them multitudinously there,
And in the hollow furnaces within
Wheel flame around, until from bursted cloud 155
In forky flashes they have gleamed forth.

Again, from following cause it comes to pass
That yon swift golden hue of liquid fire
Darts downward to the earth: because the clouds 160
Themselves must hold abundant seeds of fire;
For, when they be without all moisture, then
They be for most part of a flamy hue
And a resplendent. And, indeed, they must
Even from the light of sun unto themselves 165
Take multitudinous seeds, and so perforce
Redden and pour their bright fires all abroad.
And therefore, when the wind hath driven and thrust,
Hath forced and squeezed into one spot these clouds,
They pour abroad the seeds of fire pressed out, 170
Which make to flash these colours of the flame.
Likewise, it lightens also when the clouds
Grow rare and thin along the sky; for, when
The wind with gentle touch unravels them
And breaketh asunder as they move, those seeds 175
Which make the lightnings must by nature fall;
At such an hour the horizon lightens round
Without the hideous terror of dread noise
And skiey uproar.

180

To proceed apace,
What sort of nature thunderbolts possess
Is by their strokes made manifest and by
The brand-marks of their searing heat on things,
And by the scorched scars exhaling round 185
The heavy fumes of sulphur. For all these
Are marks, O not of wind or rain, but fire.
Again, they often enkindle even the roofs
Of houses and inside the very rooms
With swift flame hold a fierce dominion. 190
Know thou that nature fashioned this fire

Subtler than fires all other, with minute
And dartling bodies,–a fire 'gainst which there's naught
Can in the least hold out: the thunderbolt,
The mighty, passes through the hedging walls 195
Of houses, like to voices or a shout,–
Through stones, through bronze it passes, and it melts
Upon the instant bronze and gold; and makes,
Likewise, the wines sudden to vanish forth,
The wine-jars intact,–because, ye see, 200
Its heat arriving renders loose and porous
Readily all the wine–jar's earthen sides,
And winding its way within, it scattereth
The elements primordial of the wine
With speedy dissolution–process which 205
Even in an age the fiery steam of sun
Could not accomplish, however puissant he
With his hot coruscations: so much more
Agile and overpowering is this force.
210

Now in what manner engendered are these things,
How fashioned of such impetuous strength
As to cleave towers asunder, and houses all
To overtopple, and to wrench apart
Timbers and beams, and heroes' monuments 215
To pile in ruins and upheave amain,
And to take breath forever out of men,
And to o'erthrow the cattle everywhere,–
Yes, by what force the lightnings do all this,
All this and more, I will unfold to thee, 220
Nor longer keep thee in mere promises.

The bolts of thunder, then, must be conceived
As all begotten in those crasser clouds
Up-piled aloft; for, from the sky serene 225
And from the clouds of lighter density,
None are sent forth forever. That 'tis so
Beyond a doubt, fact plain to sense declares:
To wit, at such a time the densed clouds
So mass themselves through all the upper air 230
That we might think that round about all murk
Had parted forth from Acheron and filled

The mighty vaults of sky–so grievously,
As gathers thus the storm-clouds' gruesome might,
Do faces of black horror hang on high– 235
When tempest begins its thunderbolts to forge.
Besides, full often also out at sea
A blackest thunderhead, like cataract
Of pitch hurled down from heaven, and far away
Bulging with murkiness, down on the waves 240
Falls with vast uproar, and draws on amain
The darkling tempests big with thunderbolts
And hurricanes, itself the while so crammed
Tremendously with fires and winds, that even
Back on the lands the people shudder round 245
And seek for cover. Therefore, as I said,
The storm must be conceived as o'er our head
Towering most high; for never would the clouds
O'erwhelm the lands with such a massy dark,
Unless up-builded heap on lofty heap, 250
To shut the round sun off. Nor could the clouds,
As on they come, engulf with rain so vast
As thus to make the rivers overflow
And fields to float, if ether were not thus
Furnished with lofty-piled clouds. Lo, then, 255
Here be all things fulfilled with winds and fires–
Hence the long lightnings and the thunders loud.
For, verily, I've taught thee even now
How cavernous clouds hold seeds innumerable
Of fiery exhalations, and they must 260
From off the sunbeams and the heat of these
Take many still. And so, when that same wind
(Which, haply, into one region of the sky
Collects those clouds) hath pressed from out the same
The many fiery seeds, and with that fire 265
Hath at the same time inter-mixed itself,
O then and there that wind, a whirlwind now,
Deep in the belly of the cloud spins round
In narrow confines, and sharpens there inside
In glowing furnaces the thunderbolt. 270
For in a two-fold manner is that wind
Enkindled all: it trembles into heat
Both by its own velocity and by

Repeated touch of fire. Thereafter, when
The energy of wind is heated through 275
And the fierce impulse of the fire hath sped
Deeply within, O then the thunderbolt,
Now ripened, so to say, doth suddenly
Splinter the cloud, and the aroused flash
Leaps onward, lumining with forky light 280
All places round. And followeth anon
A clap so heavy that the skiey vaults,
As if asunder burst, seem from on high
To engulf the earth. Then fearfully a quake
Pervades the lands, and 'long the lofty skies 285
Run the far rumblings. For at such a time
Nigh the whole tempest quakes, shook through and through,
And roused are the roarings,–from which shock
Comes such resounding and abounding rain,
That all the murky ether seems to turn 290
Now into rain, and, as it tumbles down,
To summon the fields back to primeval floods:
So big the rains that be sent down on men
By burst of cloud and by the hurricane,
What time the thunder-clap, from burning bolt 295
That cracks the cloud, flies forth along. At times
The force of wind, excited from without,
Smiteth into a cloud already hot
With a ripe thunderbolt. And when that wind
Hath splintered that cloud, then down there cleaves forthwith 300
Yon fiery coil of flame which still we call,
Even with our fathers' word, a thunderbolt.
The same thing haps toward every other side
Whither that force hath swept. It happens, too,
That sometimes force of wind, though hurtled forth 305
Without all fire, yet in its voyage through space
Igniteth, whilst it comes along, along,–
Losing some larger bodies which cannot
Pass, like the others, through the bulks of air,–
And, scraping together out of air itself 310
Some smaller bodies, carries them along,
And these, commingling, by their flight make fire:
Much in the manner as oft a leaden ball
Grows hot upon its aery course, the while

It loseth many bodies of stark cold 315
And taketh into itself along the air
New particles of fire. It happens, too,
That force of blow itself arouses fire,
When force of wind, a-cold and hurtled forth
Without all fire, hath strook somewhere amain– 320
No marvel, because, when with terrific stroke
'Thas smitten, the elements of fiery-stuff
Can stream together from out the very wind
And, simultaneously, from out that thing
Which then and there receives the stroke: as flies 325
The fire when with the steel we hack the stone;
Nor yet, because the force of steel's a-cold,
Rush the less speedily together there
Under the stroke its seeds of radiance hot.
And therefore, thuswise must an object too 330
Be kindled by a thunderbolt, if haply
'Thas been adapt and suited to the flames.
Yet force of wind must not be rashly deemed
As altogether and entirely cold–
That force which is discharged from on high 335
With such stupendous power; but if 'tis not
Upon its course already kindled with fire,
It yet arriveth warmed and mixed with heat.

And, now, the speed and stroke of thunderbolt 340
Is so tremendous, and with glide so swift
Those thunderbolts rush on and down, because
Their roused force itself collects itself
First always in the clouds, and then prepares
For the huge effort of their going-forth; 345
Next, when the cloud no longer can retain
The increment of their fierce impetus,
Their force is pressed out, and therefore flies
With impetus so wondrous, like to shots
Hurled from the powerful Roman catapults. 350
Note, too, this force consists of elements
Both small and smooth, nor is there aught that can
With ease resist such nature. For it darts
Between and enters through the pores of things;
And so it never falters in delay 355

Despite innumerable collisions, but
Flies shooting onward with a swift elan.
Next, since by nature always every weight
Bears downward, doubled is the swiftness then
And that elan is still more wild and dread, 360
When, verily, to weight are added blows,
So that more madly and more fiercely then
The thunderbolt shakes into shivers all
That blocks its path, following on its way.
Then, too, because it comes along, along 365
With one continuing elan, it must
Take on velocity anew, anew,
Which still increases as it goes, and ever
Augments the bolt's vast powers and to the blow
Gives larger vigour; for it forces all, 370
All of the thunder's seeds of fire, to sweep
In a straight line unto one place, as 'twere,–
Casting them one by other, as they roll,
Into that onward course. Again, perchance,
In coming along, it pulls from out the air 375
Some certain bodies, which by their own blows
Enkindle its velocity. And, lo,
It comes through objects leaving them unharmed,
It goes through many things and leaves them whole,
Because the liquid fire flieth along 380
Athrough their pores. And much it does transfix,
When these primordial atoms of the bolt
Have fallen upon the atoms of these things
Precisely where the intertwined atoms
Are held together. And, further, easily 385
Brass it unbinds and quickly fuseth gold,
Because its force is so minutely made
Of tiny parts and elements so smooth
That easily they wind their way within,
And, when once in, quickly unbind all knots 390
And loosen all the bonds of union there.

And most in autumn is shaken the house of heaven,
The house so studded with the glittering stars,
And the whole earth around–most too in spring 395
When flowery times unfold themselves: for, lo,

In the cold season is there lack of fire,
And winds are scanty in the hot, and clouds
Have not so dense a bulk. But when, indeed,
The seasons of heaven are betwixt these twain, 400
The divers causes of the thunderbolt
Then all concur; for then both cold and heat
Are mixed in the cross-seas of the year,
So that a discord rises among things
And air in vast tumultuosity 405
Billows, infuriate with the fires and winds–
Of which the both are needed by the cloud
For fabrication of the thunderbolt.
For the first part of heat and last of cold
Is the time of spring; wherefore must things unlike 410
Do battle one with other, and, when mixed,
Tumultuously rage. And when rolls round
The latest heat mixed with the earliest chill–
The time which bears the name of autumn–then
Likewise fierce cold-spells wrestle with fierce heats. 415
On this account these seasons of the year
Are nominated "cross-seas."–And no marvel
If in those times the thunderbolts prevail
And storms are roused turbulent in heaven,
Since then both sides in dubious warfare rage 420
Tumultuously, the one with flames, the other
With winds and with waters mixed with winds.

This, this it is, O Memmius, to see through
The very nature of fire-fraught thunderbolt; 425
O this it is to mark by what blind force
It maketh each effect, and not, O not
To unwind Etrurian scrolls oracular,
Inquiring tokens of occult will of gods,
Even as to whence the flying flame hath come, 430
Or to which half of heaven it turns, or how
Through walled places it hath wound its way,
Or, after proving its dominion there,
How it hath speeded forth from thence amain,
Or what the thunderstroke portends of ill 435
From out high heaven. But if Jupiter
And other gods shake those refulgent vaults

With dread reverberations and hurl fire
Whither it pleases each, why smite they not
Mortals of reckless and revolting crimes, 440
That such may pant from a transpierced breast
Forth flames of the red levin–unto men
A drastic lesson?–why is rather he–
O he self-conscious of no foul offence–
Involved in flames, though innocent, and clasped 445
Up-caught in skiey whirlwind and in fire?
Nay, why, then, aim they at eternal wastes,
And spend themselves in vain?–perchance, even so
To exercise their arms and strengthen shoulders?
Why suffer they the Father's javelin 450
To be so blunted on the earth? And why
Doth he himself allow it, nor spare the same
Even for his enemies? O why most oft
Aims he at lofty places? Why behold we
Marks of his lightnings most on mountain tops? 455
Then for what reason shoots he at the sea?–
What sacrilege have waves and bulk of brine
And floating fields of foam been guilty of?
Besides, if 'tis his will that we beware
Against the lightning-stroke, why feareth he 460
To grant us power for to behold the shot?
And, contrariwise, if wills he to o'erwhelm us,
Quite off our guard, with fire, why thunders he
Off in yon quarter, so that we may shun?
Why rouseth he beforehand darkling air 465
And the far din and rumblings? And O how
Canst thou believe he shoots at one same time
Into diverse directions? Or darest thou
Contend that never hath it come to pass
That divers strokes have happened at one time? 470
But oft and often hath it come to pass,
And often still it must, that, even as showers
And rains o'er many regions fall, so too
Dart many thunderbolts at one same time.
Again, why never hurtles Jupiter 475
A bolt upon the lands nor pours abroad
Clap upon clap, when skies are cloudless all?
Or, say, doth he, so soon as ever the clouds

Have come thereunder, then into the same
Descend in person, that from thence he may 480
Near-by decide upon the stroke of shaft?
And, lastly, why, with devastating bolt
Shakes he asunder holy shrines of gods
And his own thrones of splendour, and to-breaks
The well-wrought idols of divinities, 485
And robs of glory his own images
By wound of violence?

But to return apace,
Easy it is from these same facts to know 490
In just what wise those things (which from their sort
The Greeks have named "bellows") do come down,
Discharged from on high, upon the seas.
For it haps that sometimes from the sky descends
Upon the seas a column, as if pushed, 495
Round which the surges seethe, tremendously
Aroused by puffing gusts; and whatso'er
Of ships are caught within that tumult then
Come into extreme peril, dashed along.
This haps when sometimes wind's aroused force 500
Can't burst the cloud it tries to, but down-weighs
That cloud, until 'tis like a column from sky
Upon the seas pushed downward–gradually,
As if a Somewhat from on high were shoved
By fist and nether thrust of arm, and lengthened 505
Far to the waves. And when the force of wind
Hath rived this cloud, from out the cloud it rushes
Down on the seas, and starts among the waves
A wondrous seething, for the eddying whirl
Descends and downward draws along with it 510
That cloud of ductile body. And soon as ever
'Thas shoved unto the levels of the main
That laden cloud, the whirl suddenly then
Plunges its whole self into the waters there
And rouses all the sea with monstrous roar, 515
Constraining it to seethe. It happens too
That very vortex of the wind involves
Itself in clouds, scraping from out the air
The seeds of cloud, and counterfeits, as 'twere,

The "bellows" pushed from heaven. And when this shape 520
Hath dropped upon the lands and burst apart,
It belches forth immeasurable might
Of whirlwind and of blast. Yet since 'tis formed
At most but rarely, and on land the hills
Must block its way, 'tis seen more oft out there 525
On the broad prospect of the level main
Along the free horizons.

Into being
The clouds condense, when in this upper space 530
Of the high heaven have gathered suddenly,
As round they flew, unnumbered particles–
World's rougher ones, which can, though interlinked
With scanty couplings, yet be fastened firm,
The one on other caught. These particles 535
First cause small clouds to form; and, thereupon,
These catch the one on other and swarm in a flock
And grow by their conjoining, and by winds
Are borne along, along, until collects
The tempest fury. Happens, too, the nearer 540
The mountain summits neighbour to the sky,
The more unceasingly their far crags smoke
With the thick darkness of swart cloud, because
When first the mists do form, ere ever the eyes
Can there behold them (tenuous as they be), 545
The carrier-winds will drive them up and on
Unto the topmost summits of the mountain;
And then at last it happens, when they be
In vaster throng upgathered, that they can
By this very condensation lie revealed, 550
And that at same time they are seen to surge
From very vertex of the mountain up
Into far ether. For very fact and feeling,
As we up-climb high mountains, proveth clear
That windy are those upward regions free. 555
Besides, the clothes hung-out along the shore,
When in they take the clinging moisture, prove
That nature lifts from over all the sea
Unnumbered particles. Whereby the more
'Tis manifest that many particles 560

Even from the salt upheavings of the main
Can rise together to augment the bulk
Of massed clouds. For moistures in these twain
Are near akin. Besides, from out all rivers,
As well as from the land itself, we see 565
Up-rising mists and steam, which like a breath
Are forced out from them and borne aloft,
To curtain heaven with their murk, and make,
By slow foregathering, the skiey clouds.
For, in addition, lo, the heat on high 570
Of constellated ether burdens down
Upon them, and by sort of condensation
Weaveth beneath the azure firmament
The reek of darkling cloud. It happens, too,
That hither to the skies from the Beyond 575
Do come those particles which make the clouds
And flying thunderheads. For I have taught
That this their number is innumerable
And infinite the sum of the Abyss,
And I have shown with what stupendous speed 580
Those bodies fly and how they're wont to pass
Amain through incommunicable space.
Therefore, 'tis not exceeding strange, if oft
In little time tempest and darkness cover
With bulking thunderheads hanging on high 585
The oceans and the lands, since everywhere
Through all the narrow tubes of yonder ether,
Yea, so to speak, through all the breathing-holes
Of the great upper-world encompassing,
There be for the primordial elements 590
Exits and entrances.

Now come, and how
The rainy moisture thickens into being
In the lofty clouds, and how upon the lands 595
'Tis then discharged in down-pour of large showers,
I will unfold. And first triumphantly
Will I persuade thee that up-rise together,
With clouds themselves, full many seeds of water
From out all things, and that they both increase– 600
Both clouds and water which is in the clouds–

In like proportion, as our frames increase
In like proportion with our blood, as well
As sweat or any moisture in our members.
Besides, the clouds take in from time to time 605
Much moisture risen from the broad marine,–
Whilst the winds bear them o'er the mighty sea,
Like hanging fleeces of white wool. Thuswise,
Even from all rivers is there lifted up
Moisture into the clouds. And when therein 610
The seeds of water so many in many ways
Have come together, augmented from all sides,
The close-jammed clouds then struggle to discharge
Their rain-storms for a two-fold reason: lo,
The wind's force crowds them, and the very excess 615
Of storm-clouds (massed in a vaster throng)
Giveth an urge and pressure from above
And makes the rains out-pour. Besides when, too,
The clouds are winnowed by the winds, or scattered
Smitten on top by heat of sun, they send 620
Their rainy moisture, and distil their drops,
Even as the wax, by fiery warmth on top,
Wasteth and liquefies abundantly.
But comes the violence of the bigger rains
When violently the clouds are weighted down 625
Both by their cumulated mass and by
The onset of the wind. And rains are wont
To endure awhile and to abide for long,
When many seeds of waters are aroused,
And clouds on clouds and racks on racks outstream 630
In piled layers and are borne along
From every quarter, and when all the earth
Smoking exhales her moisture. At such a time
When sun with beams amid the tempest-murk
Hath shone against the showers of black rains, 635
Then in the swart clouds there emerges bright
The radiance of the bow.

And as to things
Not mentioned here which of themselves do grow 640
Or of themselves are gendered, and all things
Which in the clouds condense to being–all,

Snow and the winds, hail and the hoar-frosts chill,
And freezing, mighty force–of lakes and pools
The mighty hardener, and mighty check 645
Which in the winter curbeth everywhere
The rivers as they go–'tis easy still,
Soon to discover and with mind to see
How they all happen, whereby gendered,
When once thou well hast understood just what 650
Functions have been vouchsafed from of old
Unto the procreant atoms of the world.
Now come, and what the law of earthquakes is
Hearken, and first of all take care to know
That the under-earth, like to the earth around us, 655
Is full of windy caverns all about;
And many a pool and many a grim abyss
She bears within her bosom, ay, and cliffs
And jagged scarps; and many a river, hid
Beneath her chine, rolls rapidly along 660
Its billows and plunging boulders. For clear fact
Requires that earth must be in every part
Alike in constitution. Therefore, earth,
With these things underneath affixed and set,
Trembleth above, jarred by big down-tumblings, 665
When time hath undermined the huge caves,
The subterranean. Yea, whole mountains fall,
And instantly from spot of that big jar
There quiver the tremors far and wide abroad.
And with good reason: since houses on the street 670
Begin to quake throughout, when jarred by a cart
Of no large weight; and, too, the furniture
Within the house up-bounds, when a paving-block
Gives either iron rim of the wheels a jolt.
It happens, too, when some prodigious bulk 675
Of age-worn soil is rolled from mountain slopes
Into tremendous pools of water dark,
That the reeling land itself is rocked about
By the water's undulations; as a basin
Sometimes won't come to rest until the fluid 680
Within it ceases to be rocked about
In random undulations.

And besides,
When subterranean winds, up-gathered there 685
In the hollow deeps, bulk forward from one spot,
And press with the big urge of mighty powers
Against the lofty grottos, then the earth
Bulks to that quarter whither push amain
The headlong winds. Then all the builded houses 690
Above ground–and the more, the higher up-reared
Unto the sky–lean ominously, careening
Into the same direction; and the beams,
Wrenched forward, over-hang, ready to go.
Yet dread men to believe that there awaits 695
The nature of the mighty world a time
Of doom and cataclysm, albeit they see
So great a bulk of lands to bulge and break!
And lest the winds blew back again, no force
Could rein things in nor hold from sure career 700
On to disaster. But now because those winds
Blow back and forth in alternation strong,
And, so to say, rallying charge again,
And then repulsed retreat, on this account
Earth oftener threatens than she brings to pass 705
Collapses dire. For to one side she leans,
Then back she sways; and after tottering
Forward, recovers then her seats of poise.
Thus, this is why whole houses rock, the roofs
More than the middle stories, middle more 710
Than lowest, and the lowest least of all.

Arises, too, this same great earth-quaking,
When wind and some prodigious force of air,
Collected from without or down within 715
The old telluric deeps, have hurled themselves
Amain into those caverns sub-terrene,
And there at first tumultuously chafe
Among the vasty grottos, borne about
In mad rotations, till their lashed force 720
Aroused out-bursts abroad, and then and there,
Riving the deep earth, makes a mighty chasm–
What once in Syrian Sidon did befall,
And once in Peloponnesian Aegium,

Twain cities which such out-break of wild air 725
And earth's convulsion, following hard upon,
O'erthrew of old. And many a walled town,
Besides, hath fall'n by such omnipotent
Convulsions on the land, and in the sea
Engulfed hath sunken many a city down 730
With all its populace. But if, indeed,
They burst not forth, yet is the very rush
Of the wild air and fury-force of wind
Then dissipated, like an ague-fit,
Through the innumerable pores of earth, 735
To set her all a-shake–even as a chill,
When it hath gone into our marrow-bones,
Sets us convulsively, despite ourselves,
A-shivering and a-shaking. Therefore, men
With two-fold terror bustle in alarm 740
Through cities to and fro: they fear the roofs
Above the head; and underfoot they dread
The caverns, lest the nature of the earth
Suddenly rend them open, and she gape,
Herself asunder, with tremendous maw, 745
And, all confounded, seek to chock it full
With her own ruins. Let men, then, go on
Feigning at will that heaven and earth shall be
Inviolable, entrusted evermore
To an eternal weal: and yet at times 750
The very force of danger here at hand
Prods them on some side with this goad of fear–
This among others–that the earth, withdrawn
Abruptly from under their feet, be hurried down,
Down into the abyss, and the Sum-of-Things 755
Be following after, utterly fordone,
Till be but wrack and wreckage of a world.

EXTRAORDINARY AND PARADOXICAL TELLURIC PHENOMENA 760

In chief, men marvel nature renders not
Bigger and bigger the bulk of ocean, since
So vast the down-rush of the waters be,
And every river out of every realm 765

Cometh thereto; and add the random rains
And flying tempests, which spatter every sea
And every land bedew; add their own springs:
Yet all of these unto the ocean's sum
Shall be but as the increase of a drop. 770
Wherefore 'tis less a marvel that the sea,
The mighty ocean, increaseth not. Besides,
Sun with his heat draws off a mighty part:
Yea, we behold that sun with burning beams
To dry our garments dripping all with wet; 775
And many a sea, and far out-spread beneath,
Do we behold. Therefore, however slight
The portion of wet that sun on any spot
Culls from the level main, he still will take
From off the waves in such a wide expanse 780
Abundantly. Then, further, also winds,
Sweeping the level waters, can bear off
A mighty part of wet, since we behold
Oft in a single night the highways dried
By winds, and soft mud crusted o'er at dawn. 785
Again, I've taught thee that the clouds bear off
Much moisture too, up-taken from the reaches
Of the mighty main, and sprinkle it about
O'er all the zones, when rain is on the lands
And winds convey the aery racks of vapour. 790
Lastly, since earth is porous through her frame,
And neighbours on the seas, girdling their shores,
The water's wet must seep into the lands
From briny ocean, as from lands it comes
Into the seas. For brine is filtered off, 795
And then the liquid stuff seeps back again
And all re-poureth at the river-heads,
Whence in fresh-water currents it returns
Over the lands, adown the channels which
Were cleft erstwhile and erstwhile bore along 800
The liquid-footed floods.

And now the cause
Whereby athrough the throat of Aetna's Mount
Such vast tornado-fires out-breathe at times, 805
I will unfold: for with no middling might

Of devastation the flamy tempest rose
And held dominion in Sicilian fields:
Drawing upon itself the upturned faces
Of neighbouring clans, what time they saw afar 810
The skiey vaults a-fume and sparkling all,
And filled their bosoms with dread anxiety
Of what new thing nature were travailing at.

In these affairs it much behooveth thee 815
To look both wide and deep, and far abroad
To peer to every quarter, that thou mayst
Remember how boundless is the Sum-of-Things,
And mark how infinitely small a part
Of the whole Sum is this one sky of ours– 820
O not so large a part as is one man
Of the whole earth. And plainly if thou viewest
This cosmic fact, placing it square in front,
And plainly understandest, thou wilt leave
Wondering at many things. For who of us 825
Wondereth if some one gets into his joints
A fever, gathering head with fiery heat,
Or any other dolorous disease
Along his members? For anon the foot
Grows blue and bulbous; often the sharp twinge 830
Seizes the teeth, attacks the very eyes;
Out-breaks the sacred fire, and, crawling on
Over the body, burneth every part
It seizeth on, and works its hideous way
Along the frame. No marvel this, since, lo, 835
Of things innumerable be seeds enough,
And this our earth and sky do bring to us
Enough of bane from whence can grow the strength
Of maladies uncounted. Thuswise, then,
We must suppose to all the sky and earth 840
Are ever supplied from out the infinite
All things, O all in stores enough whereby
The shaken earth can of a sudden move,
And fierce typhoons can over sea and lands
Go tearing on, and Aetna's fires o'erflow, 845
And heaven become a flame-burst. For that, too,
Happens at times, and the celestial vaults

Glow into fire, and rainy tempests rise
In heavier congregation, when, percase,
The seeds of water have foregathered thus 850
From out the infinite. "Aye, but passing huge
The fiery turmoil of that conflagration!"
So sayst thou; well, huge many a river seems
To him that erstwhile ne'er a larger saw;
Thus, huge seems tree or man; and everything 855
Which mortal sees the biggest of each class,
That he imagines to be "huge"; though yet
All these, with sky and land and sea to boot,
Are all as nothing to the sum entire
Of the all-Sum. 860

But now I will unfold
At last how yonder suddenly angered flame
Out-blows abroad from vasty furnaces
Aetnaean. First, the mountain's nature is 865
All under-hollow, propped about, about
With caverns of basaltic piers. And, lo,
In all its grottos be there wind and air–
For wind is made when air hath been uproused
By violent agitation. When this air 870
Is heated through and through, and, raging round,
Hath made the earth and all the rocks it touches
Horribly hot, and hath struck off from them
Fierce fire of swiftest flame, it lifts itself
And hurtles thus straight upwards through its throat 875
Into high heav'n, and thus bears on afar
Its burning blasts and scattereth afar
Its ashes, and rolls a smoke of pitchy murk
And heaveth the while boulders of wondrous weight–
Leaving no doubt in thee that 'tis the air's 880
Tumultuous power. Besides, in mighty part,
The sea there at the roots of that same mount
Breaks its old billows and sucks back its surf.
And grottos from the sea pass in below
Even to the bottom of the mountain's throat. 885
Herethrough thou must admit there go...

And the conditions force [the water and air]

Deeply to penetrate from the open sea,
And to out-blow abroad, and to up-bear 890
Thereby the flame, and to up-cast from deeps
The boulders, and to rear the clouds of sand.
For at the top be "bowls," as people there
Are wont to name what we at Rome do call
The throats and mouths. 895

There be, besides, some thing
Of which 'tis not enough one only cause
To state–but rather several, whereof one
Will be the true: lo, if thou shouldst espy 900
Lying afar some fellow's lifeless corse,
'Twere meet to name all causes of a death,
That cause of his death might thereby be named:
For prove thou mayst he perished not by steel,
By cold, nor even by poison nor disease, 905
Yet somewhat of this sort hath come to him
We know–And thus we have to say the same
In divers cases.

Toward the summer, Nile 910
Waxeth and overfloweth the champaign,
Unique in all the landscape, river sole
Of the Aegyptians. In mid-season heats
Often and oft he waters Aegypt o'er,
Either because in summer against his mouths 915
Come those northwinds which at that time of year
Men name the Etesian blasts, and, blowing thus
Upstream, retard, and, forcing back his waves,
Fill him o'erfull and force his flow to stop.
For out of doubt these blasts which driven be 920
From icy constellations of the pole
Are borne straight up the river. Comes that river
From forth the sultry places down the south,
Rising far up in midmost realm of day,
Among black generations of strong men 925
With sun-baked skins. 'Tis possible, besides,
That a big bulk of piled sand may bar
His mouths against his onward waves, when sea,
Wild in the winds, tumbles the sand to inland;

Whereby the river's outlet were less free, 930
Likewise less headlong his descending floods.
It may be, too, that in this season rains
Are more abundant at its fountain head,
Because the Etesian blasts of those northwinds
Then urge all clouds into those inland parts. 935
And, soothly, when they're thus foregathered there,
Urged yonder into midmost realm of day,
Then, crowded against the lofty mountain sides,
They're massed and powerfully pressed. Again,
Perchance, his waters wax, O far away, 940
Among the Aethiopians' lofty mountains,
When the all-beholding sun with thawing beams
Drives the white snows to flow into the vales.

Now come; and unto thee I will unfold, 945
As to the Birdless spots and Birdless tarns,
What sort of nature they are furnished with.
First, as to name of "birdless,"–that derives
From very fact, because they noxious be
Unto all birds. For when above those spots 950
In horizontal flight the birds have come,
Forgetting to oar with wings, they furl their sails,
And, with down-drooping of their delicate necks,
Fall headlong into earth, if haply such
The nature of the spots, or into water, 955
If haply spreads thereunder Birdless tarn.
Such spot's at Cumae, where the mountains smoke,
Charged with the pungent sulphur, and increased
With steaming springs. And such a spot there is
Within the walls of Athens, even there 960
On summit of Acropolis, beside
Fane of Tritonian Pallas bountiful,
Where never cawing crows can wing their course,
Not even when smoke the altars with good gifts,–
But evermore they flee–yet not from wrath 965
Of Pallas, grieved at that espial old,
As poets of the Greeks have sung the tale;
But very nature of the place compels.
In Syria also–as men say–a spot
Is to be seen, where also four-foot kinds, 970

As soon as ever they've set their steps within,
Collapse, o'ercome by its essential power,
As if there slaughtered to the under-gods.
Lo, all these wonders work by natural law,
And from what causes they are brought to pass 975
The origin is manifest; so, haply,
Let none believe that in these regions stands
The gate of Orcus, nor us then suppose,
Haply, that thence the under-gods draw down
Souls to dark shores of Acheron–as stags, 980
The wing-footed, are thought to draw to light,
By sniffing nostrils, from their dusky lairs
The wriggling generations of wild snakes.
How far removed from true reason is this,
Perceive thou straight; for now I'll try to say 985
Somewhat about the very fact.

And, first,
This do I say, as oft I've said before:
In earth are atoms of things of every sort; 990
And know, these all thus rise from out the earth–
Many life-giving which be good for food,
And many which can generate disease
And hasten death, O many primal seeds
Of many things in many modes–since earth 995
Contains them mingled and gives forth discrete.
And we have shown before that certain things
Be unto certain creatures suited more
For ends of life, by virtue of a nature,
A texture, and primordial shapes, unlike 1000
For kinds alike. Then too 'tis thine to see
How many things oppressive be and foul
To man, and to sensation most malign:
Many meander miserably through ears;
Many in-wind athrough the nostrils too, 1005
Malign and harsh when mortal draws a breath;
Of not a few must one avoid the touch;
Of not a few must one escape the sight;
And some there be all loathsome to the taste;
And many, besides, relax the languid limbs 1010
Along the frame, and undermine the soul

In its abodes within. To certain trees
There hath been given so dolorous a shade
That often they gender achings of the head,
If one but be beneath, outstretched on the sward. 1015
There is, again, on Helicon's high hills
A tree that's wont to kill a man outright
By fetid odour of its very flower.
And when the pungent stench of the night-lamp,
Extinguished but a moment since, assails 1020
The nostrils, then and there it puts to sleep
A man afflicted with the falling sickness
And foamings at the mouth. A woman, too,
At the heavy castor drowses back in chair,
And from her delicate fingers slips away 1025
Her gaudy handiwork, if haply she
Hath got the whiff at menstruation-time.
Once more, if thou delayest in hot baths,
When thou art over-full, how readily
From stool in middle of the steaming water 1030
Thou tumblest in a fit! How readily
The heavy fumes of charcoal wind their way
Into the brain, unless beforehand we
Of water 've drunk. But when a burning fever,
O'ermastering man, hath seized upon his limbs, 1035
Then odour of wine is like a hammer-blow.
And seest thou not how in the very earth
Sulphur is gendered and bitumen thickens
With noisome stench?–What direful stenches, too,
Scaptensula out-breathes from down below, 1040
When men pursue the veins of silver and gold,
With pick-axe probing round the hidden realms
Deep in the earth?–Or what of deadly bane
The mines of gold exhale? O what a look,
And what a ghastly hue they give to men! 1045
And seest thou not, or hearest, how they're wont
In little time to perish, and how fail
The life-stores in those folk whom mighty power
Of grim necessity confineth there
In such a task? Thus, this telluric earth 1050
Out-streams with all these dread effluvia
And breathes them out into the open world

And into the visible regions under heaven.

Thus, too, those Birdless places must up-send 1055
An essence bearing death to winged things,
Which from the earth rises into the breezes
To poison part of skiey space, and when
Thither the winged is on pennons borne,
There, seized by the unseen poison, 'tis ensnared, 1060
And from the horizontal of its flight
Drops to the spot whence sprang the effluvium.
And when 'thas there collapsed, then the same power
Of that effluvium takes from all its limbs
The relics of its life. That power first strikes 1065
The creatures with a wildering dizziness,
And then thereafter, when they're once down-fallen
Into the poison's very fountains, then
Life, too, they vomit out perforce, because
So thick the stores of bane around them fume. 1070

Again, at times it happens that this power,
This exhalation of the Birdless places,
Dispels the air betwixt the ground and birds,
Leaving well-nigh a void. And thither when 1075
In horizontal flight the birds have come,
Forthwith their buoyancy of pennons limps,
All useless, and each effort of both wings
Falls out in vain. Here, when without all power
To buoy themselves and on their wings to lean, 1080
Lo, nature constrains them by their weight to slip
Down to the earth, and lying prostrate there
Along the well-nigh empty void, they spend
Their souls through all the openings of their frame.
1085

Further, the water of wells is colder then
At summer time, because the earth by heat
Is rarefied, and sends abroad in air
Whatever seeds it peradventure have
Of its own fiery exhalations. 1090
The more, then, the telluric ground is drained
Of heat, the colder grows the water hid
Within the earth. Further, when all the earth

Is by the cold compressed, and thus contracts
And, so to say, concretes, it happens, lo, 1095
That by contracting it expresses then
Into the wells what heat it bears itself.

'Tis said at Hammon's fane a fountain is,
In daylight cold and hot in time of night. 1100
This fountain men be-wonder over-much,
And think that suddenly it seethes in heat
By intense sun, the subterranean, when
Night with her terrible murk hath cloaked the lands–
What's not true reasoning by a long remove: 1105
I' faith when sun o'erhead, touching with beams
An open body of water, had no power
To render it hot upon its upper side,
Though his high light possess such burning glare,
How, then, can he, when under the gross earth, 1110
Make water boil and glut with fiery heat?–
And, specially, since scarcely potent he
Through hedging walls of houses to inject
His exhalations hot, with ardent rays.
What, then's, the principle? Why, this, indeed: 1115
The earth about that spring is porous more
Than elsewhere the telluric ground, and be
Many the seeds of fire hard by the water;
On this account, when night with dew-fraught shades
Hath whelmed the earth, anon the earth deep down 1120
Grows chill, contracts; and thuswise squeezes out
Into the spring what seeds she holds of fire
(As one might squeeze with fist), which render hot
The touch and steam of the fluid. Next, when sun,
Up-risen, with his rays has split the soil 1125
And rarefied the earth with waxing heat,
Again into their ancient abodes return
The seeds of fire, and all the Hot of water
Into the earth retires; and this is why
The fountain in the daylight gets so cold. 1130
Besides, the water's wet is beat upon
By rays of sun, and, with the dawn, becomes
Rarer in texture under his pulsing blaze;
And, therefore, whatso seeds it holds of fire

It renders up, even as it renders oft 1135
The frost that it contains within itself
And thaws its ice and looseneth the knots.
There is, moreover, a fountain cold in kind
That makes a bit of tow (above it held)
Take fire forthwith and shoot a flame; so, too, 1140
A pitch-pine torch will kindle and flare round
Along its waves, wherever 'tis impelled
Afloat before the breeze. No marvel, this:
Because full many seeds of heat there be
Within the water; and, from earth itself 1145
Out of the deeps must particles of fire
Athrough the entire fountain surge aloft,
And speed in exhalations into air
Forth and abroad (yet not in numbers enow
As to make hot the fountain). And, moreo'er, 1150
Some force constrains them, scattered through the water,
Forthwith to burst abroad, and to combine
In flame above. Even as a fountain far
There is at Aradus amid the sea,
Which bubbles out sweet water and disparts 1155
From round itself the salt waves; and, behold,
In many another region the broad main
Yields to the thirsty mariners timely help,
Belching sweet waters forth amid salt waves.
Just so, then, can those seeds of fire burst forth 1160
Athrough that other fount, and bubble out
Abroad against the bit of tow; and when
They there collect or cleave unto the torch,
Forthwith they readily flash aflame, because
The tow and torches, also, in themselves 1165
Have many seeds of latent fire. Indeed,
And seest thou not, when near the nightly lamps
Thou bringest a flaxen wick, extinguished
A moment since, it catches fire before
'Thas touched the flame, and in same wise a torch? 1170
And many another object flashes aflame
When at a distance, touched by heat alone,
Before 'tis steeped in veritable fire.
This, then, we must suppose to come to pass
In that spring also. 1175

Now to other things!
And I'll begin to treat by what decree
Of nature it came to pass that iron can be
By that stone drawn which Greeks the magnet call
After the country's name (its origin 1180
Being in country of Magnesian folk).
This stone men marvel at; and sure it oft
Maketh a chain of rings, depending, lo,
From off itself! Nay, thou mayest see at times
Five or yet more in order dangling down 1185
And swaying in the delicate winds, whilst one
Depends from other, cleaving to under-side,
And ilk one feels the stone's own power and bonds–
So over-masteringly its power flows down.

1190

In things of this sort, much must be made sure
Ere thou account of the thing itself canst give,
And the approaches roundabout must be;
Wherefore the more do I exact of thee
A mind and ears attent. 1195

First, from all things
We see soever, evermore must flow,
Must be discharged and strewn about, about,
Bodies that strike the eyes, awaking sight. 1200
From certain things flow odours evermore,
As cold from rivers, heat from sun, and spray
From waves of ocean, eater-out of walls
Along the coasts. Nor ever cease to seep
The varied echoings athrough the air. 1205
Then, too, there comes into the mouth at times
The wet of a salt taste, when by the sea
We roam about; and so, whene'er we watch
The wormwood being mixed, its bitter stings.
To such degree from all things is each thing 1210
Borne streamingly along, and sent about
To every region round; and nature grants
Nor rest nor respite of the onward flow,
Since 'tis incessantly we feeling have,
And all the time are suffered to descry 1215
And smell all things at hand, and hear them sound.

Now will I seek again to bring to mind
How porous a body all things have–a fact
Made manifest in my first canto, too.
For, truly, though to know this doth import 1220
For many things, yet for this very thing
On which straightway I'm going to discourse,
'Tis needful most of all to make it sure
That naught's at hand but body mixed with void.
A first ensample: in grottos, rocks o'erhead 1225
Sweat moisture and distil the oozy drops;
Likewise, from all our body seeps the sweat;
There grows the beard, and along our members all
And along our frame the hairs. Through all our veins
Disseminates the foods, and gives increase 1230
And aliment down to the extreme parts,
Even to the tiniest finger-nails. Likewise,
Through solid bronze the cold and fiery heat
We feel to pass; likewise, we feel them pass
Through gold, through silver, when we clasp in hand 1235
The brimming goblets. And, again, there flit
Voices through houses' hedging walls of stone;
Odour seeps through, and cold, and heat of fire
That's wont to penetrate even strength of iron.
Again, where corselet of the sky girds round 1240

And at same time, some Influence of bane,
When from Beyond 'thas stolen into [our world].
And tempests, gathering from the earth and sky,
Back to the sky and earth absorbed retire– 1245
With reason, since there's naught that's fashioned not
With body porous.

Furthermore, not all
The particles which be from things thrown off 1250
Are furnished with same qualities for sense,
Nor be for all things equally adapt.
A first ensample: the sun doth bake and parch
The earth; but ice he thaws, and with his beams
Compels the lofty snows, up-reared white 1255
Upon the lofty hills, to waste away;
Then, wax, if set beneath the heat of him,

Melts to a liquid. And the fire, likewise,
Will melt the copper and will fuse the gold,
But hides and flesh it shrivels up and shrinks. 1260
The water hardens the iron just off the fire,
But hides and flesh (made hard by heat) it softens.
The oleaster-tree as much delights
The bearded she-goats, verily as though
'Twere nectar-steeped and shed ambrosia; 1265
Than which is naught that burgeons into leaf
More bitter food for man. A hog draws back
For marjoram oil, and every unguent fears
Fierce poison these unto the bristled hogs,
Yet unto us from time to time they seem, 1270
As 'twere, to give new life. But, contrariwise,
Though unto us the mire be filth most foul,
To hogs that mire doth so delightsome seem
That they with wallowing from belly to back
Are never cloyed. 1275

A point remains, besides,
Which best it seems to tell of, ere I go
To telling of the fact at hand itself.
Since to the varied things assigned be 1280
The many pores, those pores must be diverse
In nature one from other, and each have
Its very shape, its own direction fixed.
And so, indeed, in breathing creatures be
The several senses, of which each takes in 1285
Unto itself, in its own fashion ever,
Its own peculiar object. For we mark
How sounds do into one place penetrate,
Into another flavours of all juice,
And savour of smell into a third. Moreover, 1290
One sort through rocks we see to seep, and, lo,
One sort to pass through wood, another still
Through gold, and others to go out and off
Through silver and through glass. For we do see
Through some pores form-and-look of things to flow, 1295
Through others heat to go, and some things still
To speedier pass than others through same pores.
Of verity, the nature of these same paths,

Varying in many modes (as aforesaid)
Because of unlike nature and warp and woof 1300
Of cosmic things, constrains it so to be.

Wherefore, since all these matters now have been
Established and settled well for us
As premises prepared, for what remains 1305
'Twill not be hard to render clear account
By means of these, and the whole cause reveal
Whereby the magnet lures the strength of iron.
First, stream there must from off the lode-stone seeds
Innumerable, a very tide, which smites 1310
By blows that air asunder lying betwixt
The stone and iron. And when is emptied out
This space, and a large place between the two
Is made a void, forthwith the primal germs
Of iron, headlong slipping, fall conjoined 1315
Into the vacuum, and the ring itself
By reason thereof doth follow after and go
Thuswise with all its body. And naught there is
That of its own primordial elements
More thoroughly knit or tighter linked coheres 1320
Than nature and cold roughness of stout iron.
Wherefore, 'tis less a marvel what I said,
That from such elements no bodies can
From out the iron collect in larger throng
And be into the vacuum borne along, 1325
Without the ring itself do follow after.
And this it does, and followeth on until
'Thath reached the stone itself and cleaved to it
By links invisible. Moreover, likewise,
The motion's assisted by a thing of aid 1330
(Whereby the process easier becomes),–
Namely, by this: as soon as rarer grows
That air in front of the ring, and space between
Is emptied more and made a void, forthwith
It happens all the air that lies behind 1335
Conveys it onward, pushing from the rear.
For ever doth the circumambient air
Drub things unmoved, but here it pushes forth
The iron, because upon one side the space

Lies void and thus receives the iron in. 1340
This air, whereof I am reminding thee,
Winding athrough the iron's abundant pores
So subtly into the tiny parts thereof,
Shoves it and pushes, as wind the ship and sails.
The same doth happen in all directions forth: 1345
From whatso side a space is made a void,
Whether from crosswise or above, forthwith
The neighbour particles are borne along
Into the vacuum; for of verity,
They're set a-going by poundings from elsewhere, 1350
Nor by themselves of own accord can they
Rise upwards into the air. Again, all things
Must in their framework hold some air, because
They are of framework porous, and the air
Encompasses and borders on all things. 1355
Thus, then, this air in iron so deeply stored
Is tossed evermore in vexed motion,
And therefore drubs upon the ring sans doubt
And shakes it up inside....

1360

In sooth, that ring is thither borne along
To where 'thas once plunged headlong–thither, lo,
Unto the void whereto it took its start.

It happens, too, at times that nature of iron 1365
Shrinks from this stone away, accustomed
By turns to flee and follow. Yea, I've seen
Those Samothracian iron rings leap up,
And iron filings in the brazen bowls
Seethe furiously, when underneath was set 1370
The magnet stone. So strongly iron seems
To crave to flee that rock. Such discord great
Is gendered by the interposed brass,
Because, forsooth, when first the tide of brass
Hath seized upon and held possession of 1375
The iron's open passage-ways, thereafter
Cometh the tide of the stone, and in that iron
Findeth all spaces full, nor now hath holes
To swim through, as before. 'Tis thus constrained
With its own current 'gainst the iron's fabric 1380

To dash and beat; by means whereof it spues
Forth from itself–and through the brass stirs up–
The things which otherwise without the brass
It sucks into itself. In these affairs
Marvel thou not that from this stone the tide 1385
Prevails not likewise other things to move
With its own blows: for some stand firm by weight,
As gold; and some cannot be moved forever,
Because so porous in their framework they
That there the tide streams through without a break, 1390
Of which sort stuff of wood is seen to be.
Therefore, when iron (which lies between the two)
Hath taken in some atoms of the brass,
Then do the streams of that Magnesian rock
Move iron by their smitings. 1395

Yet these things
Are not so alien from others, that I
Of this same sort am ill prepared to name
Ensamples still of things exclusively 1400
To one another adapt. Thou seest, first,
How lime alone cementeth stones: how wood
Only by glue-of-bull with wood is joined–
So firmly too that oftener the boards
Crack open along the weakness of the grain 1405
Ere ever those taurine bonds will lax their hold.
The vine-born juices with the water-springs
Are bold to mix, though not the heavy pitch
With the light oil-of-olive. And purple dye
Of shell-fish so uniteth with the wool's 1410
Body alone that it cannot be ta'en
Away forever–nay, though thou gavest toil
To restore the same with the Neptunian flood,
Nay, though all ocean willed to wash it out
With all its waves. Again, gold unto gold 1415
Doth not one substance bind, and only one?
And is not brass by tin joined unto brass?
And other ensamples how many might one find!
What then? Nor is there unto thee a need
Of such long ways and roundabout, nor boots it 1420
For me much toil on this to spend. More fit

It is in few words briefly to embrace
Things many: things whose textures fall together
So mutually adapt, that cavities
To solids correspond, these cavities 1425
Of this thing to the solid parts of that,
And those of that to solid parts of this–
Such joinings are the best. Again, some things
Can be the one with other coupled and held,
Linked by hooks and eyes, as 'twere; and this 1430
Seems more the fact with iron and this stone.
Now, of diseases what the law, and whence
The Influence of bane upgathering can
Upon the race of man and herds of cattle
Kindle a devastation fraught with death, 1435
I will unfold. And, first, I've taught above
That seeds there be of many things to us
Life-giving, and that, contrariwise, there must
Fly many round bringing disease and death.
When these have, haply, chanced to collect 1440
And to derange the atmosphere of earth,
The air becometh baneful. And, lo, all
That Influence of bane, that pestilence,
Or from Beyond down through our atmosphere,
Like clouds and mists, descends, or else collects 1445
From earth herself and rises, when, a-soak
And beat by rains unseasonable and suns,
Our earth hath then contracted stench and rot.
Seest thou not, also, that whoso arrive
In region far from fatherland and home 1450
Are by the strangeness of the clime and waters
Distempered?–since conditions vary much.
For in what else may we suppose the clime
Among the Britons to differ from Aegypt's own
(Where totters awry the axis of the world), 1455
Or in what else to differ Pontic clime
From Gades' and from climes adown the south,
On to black generations of strong men
With sun-baked skins? Even as we thus do see
Four climes diverse under the four main-winds 1460
And under the four main-regions of the sky,
So, too, are seen the colour and face of men

Vastly to disagree, and fixed diseases
To seize the generations, kind by kind:
There is the elephant-disease which down 1465
In midmost Aegypt, hard by streams of Nile,
Engendered is–and never otherwhere.
In Attica the feet are oft attacked,
And in Achaean lands the eyes. And so
The divers spots to divers parts and limbs 1470
Are noxious; 'tis a variable air
That causes this. Thus when an atmosphere,
Alien by chance to us, begins to heave,
And noxious airs begin to crawl along,
They creep and wind like unto mist and cloud, 1475
Slowly, and everything upon their way
They disarrange and force to change its state.
It happens, too, that when they've come at last
Into this atmosphere of ours, they taint
And make it like themselves and alien. 1480
Therefore, asudden this devastation strange,
This pestilence, upon the waters falls,
Or settles on the very crops of grain
Or other meat of men and feed of flocks.
Or it remains a subtle force, suspense 1485
In the atmosphere itself; and when therefrom
We draw our inhalations of mixed air,
Into our body equally its bane
Also we must suck in. In manner like,
Oft comes the pestilence upon the kine, 1490
And sickness, too, upon the sluggish sheep.
Nor aught it matters whether journey we
To regions adverse to ourselves and change
The atmospheric cloak, or whether nature
Herself import a tainted atmosphere 1495
To us or something strange to our own use
Which can attack us soon as ever it come.

The Plague Athens

'Twas such a manner of disease, 'twas such
Mortal miasma in Cecropian lands
Whilom reduced the plains to dead men's bones,
Unpeopled the highways, drained of citizens
The Athenian town. For coming from afar, 5
Rising in lands of Aegypt, traversing
Reaches of air and floating fields of foam,
At last on all Pandion's folk it swooped;
Whereat by troops unto disease and death
Were they o'er-given. At first, they'd bear about 10
A skull on fire with heat, and eyeballs twain
Red with suffusion of blank glare. Their throats,
Black on the inside, sweated oozy blood;
And the walled pathway of the voice of man
Was clogged with ulcers; and the very tongue, 15
The mind's interpreter, would trickle gore,
Weakened by torments, tardy, rough to touch.
Next when that Influence of bane had chocked,
Down through the throat, the breast, and streamed had
E'en into sullen heart of those sick folk, 20
Then, verily, all the fences of man's life
Began to topple. From the mouth the breath
Would roll a noisome stink, as stink to heaven
Rotting cadavers flung unburied out.
And, lo, thereafter, all the body's strength 25
And every power of mind would languish, now
In very doorway of destruction.
And anxious anguish and ululation (mixed
With many a groan) companioned alway

The intolerable torments. Night and day, 30
Recurrent spasms of vomiting would rack
Alway their thews and members, breaking down
With sheer exhaustion men already spent.
And yet on no one's body couldst thou mark
The skin with o'er-much heat to burn aglow, 35
But rather the body unto touch of hands
Would offer a warmish feeling, and thereby
Show red all over, with ulcers, so to say,
Inbranded, like the "sacred fires" o'erspread
Along the members. The inward parts of men, 40
In truth, would blaze unto the very bones;
A flame, like flame in furnaces, would blaze
Within the stomach. Nor couldst aught apply
Unto their members light enough and thin
For shift of aid–but coolness and a breeze 45
Ever and ever. Some would plunge those limbs
On fire with bane into the icy streams,
Hurling the body naked into the waves;
Many would headlong fling them deeply down
The water-pits, tumbling with eager mouth 50
Already agape. The insatiable thirst
That whelmed their parched bodies, lo, would make
A goodly shower seem like to scanty drops.
Respite of torment was there none. Their frames
Forspent lay prone. With silent lips of fear 55
Would Medicine mumble low, the while she saw
So many a time men roll their eyeballs round,
Staring wide-open, unvisited of sleep,
The heralds of old death. And in those months
Was given many another sign of death: 60
The intellect of mind by sorrow and dread
Deranged, the sad brow, the countenance
Fierce and delirious, the tormented ears
Beset with ringings, the breath quick and short
Or huge and intermittent, soaking sweat 65
A-glisten on neck, the spittle in fine gouts
Tainted with colour of crocus and so salt,
The cough scarce wheezing through the rattling throat.
Aye, and the sinews in the fingered hands
Were sure to contract, and sure the jointed frame 70

To shiver, and up from feet the cold to mount
Inch after inch: and toward the supreme hour
At last the pinched nostrils, nose's tip
A very point, eyes sunken, temples hollow,
Skin cold and hard, the shuddering grimace, 75
The pulled and puffy flesh above the brows!–
O not long after would their frames lie prone
In rigid death. And by about the eighth
Resplendent light of sun, or at the most
On the ninth flaming of his flambeau, they 80
Would render up the life. If any then
Had 'scaped the doom of that destruction, yet
Him there awaited in the after days
A wasting and a death from ulcers vile
And black discharges of the belly, or else 85
Through the clogged nostrils would there ooze along
Much fouled blood, oft with an aching head:
Hither would stream a man's whole strength and flesh.
And whoso had survived that virulent flow
Of the vile blood, yet into thews of him 90
And into his joints and very genitals
Would pass the old disease. And some there were,
Dreading the doorways of destruction
So much, lived on, deprived by the knife
Of the male member; not a few, though lopped 95
Of hands and feet, would yet persist in life,
And some there were who lost their eyeballs: O
So fierce a fear of death had fallen on them!
And some, besides, were by oblivion
Of all things seized, that even themselves they knew 100
No longer. And though corpse on corpse lay piled
Unburied on ground, the race of birds and beasts
Would or spring back, scurrying to escape
The virulent stench, or, if they'd tasted there,
Would languish in approaching death. But yet 105
Hardly at all during those many suns
Appeared a fowl, nor from the woods went forth
The sullen generations of wild beasts–
They languished with disease and died and died.
In chief, the faithful dogs, in all the streets 110
Outstretched, would yield their breath distressfully

For so that Influence of bane would twist
Life from their members. Nor was found one sure
And universal principle of cure:
For what to one had given the power to take 115
The vital winds of air into his mouth,
And to gaze upward at the vaults of sky,
The same to others was their death and doom.

In those affairs, O awfullest of all, 120
O pitiable most was this, was this:
Whoso once saw himself in that disease
Entangled, ay, as damned unto death,
Would lie in wanhope, with a sullen heart,
Would, in fore-vision of his funeral, 125
Give up the ghost, O then and there. For, lo,
At no time did they cease one from another
To catch contagion of the greedy plague,–
As though but woolly flocks and horned herds;
And this in chief would heap the dead on dead: 130
For who forbore to look to their own sick,
O these (too eager of life, of death afeard)
Would then, soon after, slaughtering Neglect
Visit with vengeance of evil death and base–
Themselves deserted and forlorn of help. 135
But who had stayed at hand would perish there
By that contagion and the toil which then
A sense of honour and the pleading voice
Of weary watchers, mixed with voice of wail
Of dying folk, forced them to undergo. 140
This kind of death each nobler soul would meet.
The funerals, uncompanioned, forsaken,
Like rivals contended to be hurried through.

And men contending to ensepulchre 145
Pile upon pile the throng of their own dead:
And weary with woe and weeping wandered home;
And then the most would take to bed from grief.
Nor could be found not one, whom nor disease
Nor death, nor woe had not in those dread times 150
Attacked.

By now the shepherds and neatherds all,
Yea, even the sturdy guiders of curved ploughs,
Began to sicken, and their bodies would lie 155
Huddled within back-corners of their huts,
Delivered by squalor and disease to death.
O often and often couldst thou then have seen
On lifeless children lifeless parents prone,
Or offspring on their fathers', mothers' corpse 160
Yielding the life. And into the city poured
O not in least part from the countryside
That tribulation, which the peasantry
Sick, sick, brought thither, thronging from every quarter,
Plague-stricken mob. All places would they crowd, 165
All buildings too; whereby the more would death
Up-pile a-heap the folk so crammed in town.
Ah, many a body thirst had dragged and rolled
Along the highways there was lying strewn
Besides Silenus-headed water-fountains,– 170
The life-breath choked from that too dear desire
Of pleasant waters. Ah, everywhere along
The open places of the populace,
And along the highways, O thou mightest see
Of many a half-dead body the sagged limbs, 175
Rough with squalor, wrapped around with rags,
Perish from very nastiness, with naught
But skin upon the bones, well-nigh already
Buried–in ulcers vile and obscene filth.
All holy temples, too, of deities 180
Had Death becrammed with the carcasses;
And stood each fane of the Celestial Ones
Laden with stark cadavers everywhere–
Places which warders of the shrines had crowded
With many a guest. For now no longer men 185
Did mightily esteem the old Divine,
The worship of the gods: the woe at hand
Did over-master. Nor in the city then
Remained those rites of sepulture, with which
That pious folk had evermore been wont 190
To buried be. For it was wildered all
In wild alarms, and each and every one
With sullen sorrow would bury his own dead,

As present shift allowed. And sudden stress
And poverty to many an awful act 195
Impelled; and with a monstrous screaming they
Would, on the frames of alien funeral pyres,
Place their own kin, and thrust the torch beneath
Oft brawling with much bloodshed round about
Rather than quit dead bodies loved in life. 200

20,000 Leagues Under the Sea BY JULES VERNE
A Christmas Carol BY CHARLES DICKENS
A Fighting Man of Mars BY EDGAR RICE BURROUGHS
A Portrait of the Artist as a Young Man BY JAMES JOYCE
A Princess of Mars BY EDGAR RICE BURROUGHS
A Room with a View BY E. M. FORSTER
A Study in Scarlet BY ARTHUR CONAN DOYLE
A Tale of Two Cities BY JULES VERNE
Aesop's Fables BY AESOP
Alice in Wonderland BY LEWIS CARROLL
Anna Karenina BY LEO TOLSTOY
Anne of Avonlea BY LUCY MAUD MONTGOMERY
Anne of Green Gables BY LUCY MAUD MONTGOMERY
Around the World in 80 Days BY JULES VERNE
As a Man Thinketh BY JAMES ALLEN
Black Beauty BY ANNA SEWELL
Bleak House BY CHARLES DICKENS
Candide BY VOLTAIRE
Common Sense BY THOMAS PAINE
Crime and Punishment BY FYODOR DOSTOEVSKY
Don Quixote BY MIGUEL DE CERVANTES
Dracula BY BRAM STOKER
Dubliners BY JAMES JOYCE
Emily of New Moon BY LUCY MAUD MONTGOMERY
Emma BY JANE AUSTEN
Far from the Madding Crowd BY THOMAS HARDY
Frankenstein BY MARY SHELLEY
From the Earth to the Moon BY JULES VERNE
Great Expectations BY CHARLES DICKENS
Grimm's Fairy Tales BY JAKOB AND WILHELM GRIMM
Gulliver's Travels BY JONATHAN SWIFT
Hamlet BY WILLIAM SHAKESPEARE
Hard Times BY CHARLES DICKENS
Heart of Darkness BY JOSEPH CONRAD
Howards End BY E. M. FORSTER
Jane Eyre BY CHARLOTTE BRONTË
Journey to the Center of the Earth BY JULES VERNE
Kidnapped BY ROBERT LOUIS STEVENSON
Kim BY RUDYARD KIPLING
Les Misérables BY VICTOR HUGO
Leviathan BY THOMAS HOBBES
Little Men BY LOUISA MAY ALCOTT

Little Women BY LOUISA MAY ALCOTT
Macbeth BY WILLIAM SHAKESPEARE
Madame Bovary BY GUSTAVE FLAUBERT
Mansfield Park BY JANE AUSTEN
Meditations BY MARCUS AURELIUS
Middlemarch BY GEORGE ELIOT
Moby Dick BY HERMAN MELVILLE
Moll Flanders BY DANIEL DEFOE
My Ántonia BY WILLA CATHER
Nicomachean Ethics BY ARISTOTLE
North and South BY ELIZABETH GASKELL
Northanger Abbey BY JANE AUSTEN
Notes from the Underground BY FYODOR DOSTOYEVSKY
Oliver Twist BY CHARLES DICKENS
On Liberty BY JOHN STUART MILL
On War BY CARL VON CLAUSEWITZ
Orthodoxy BY GILBERT K. CHESTERTON
Paradise Lost BY JOHN MILTON
Persuasion BY JANE AUSTEN
Plato: Five Dialogues BY PLATO
Plutarch's Lives BY PLUTARCH
Politics BY ARISTOTLE
Pollyanna BY ELEANOR H. PORTER
Pride & Prejudice BY JANE AUSTEN
Raggedy Ann Stories BY JOHNNY GRUELLE
Robinson Crusoe BY DANIEL DEFOE
Second Treatise Of Government BY JOHN LOCKE
Self-Reliance, Nature, and Other Essays BY RALPH WALDO EMERSON
Sense & Sensibility BY JANE AUSTEN
Sons and Lovers BY D. H. LAWRENCE
Swann's Way, In Search of Lost Time BY MARCEL PROUST
Tarzan of the Apes BY EDGAR RICE BURROUGHS
Tess of the d'Urbervilles BY THOMAS HARDY
The Adventures of Huckleberry Finn BY MARK TWAIN
The Adventures of Sherlock Holmes BY ARTHUR CONAN DOYLE
The Adventures of Tom Sawyer BY MARK TWAIN
The Age of Innocence BY EDITH WHARTON
The Art of War BY SUN TZU
The Awakening BY KATE CHOPIN
The Beautiful and the Damned BY F. SCOTT FITZGERALD
The Brothers Karamazov BY FYODOR DOSTOEVSKY
The Call of the Wild BY JACK LONDON

The Chessmen of Mars BY EDGAR RICE BURROUGHS
The Communist Manifesto BY KARL MARX & FRIEDRICH ENGELS
The Corpus Hermeticum BY HERMES TRISMEGISTUS
The Count of Monte Cristo BY ALEXANDRE DUMAS
The Divine Comedy - Inferno BY DANTE
The Divine Comedy - Purgatorio BY DANTE
The Divine Comedy - Paradiso BY DANTE
The Everlasting Man BY GILBERT K. CHESTERTON
The First Men in the Moon BY H. G. WELLS
The Gods of Mars BY EDGAR RICE BURROUGHS
The Histories BY HERODOTUS
The History of the Peloponnesian War BY THUCYDIDES
The Hound of the Baskervilles BY ARTHUR CONAN DOYLE
The Hunchback of Notre-Dame BY VICTOR HUGO
The Idiot BY FYODOR DOSTOEVSKY
The Iliad BY HOMER
The Importance of Being Earnest BY OSCAR WILDE
The Innocents Abroad BY MARK TWAIN
The Invisible Man BY H. G. WELLS
The Island of Doctor Moreau BY H. G. WELLS
The Jungle BY UPTON SINCLAIR
The Jungle Book BY RUDYARD KIPLING
The Kama Sutra BY VĀTSYĀYANA
The Kybalion BY THREE INITIATES
The Last of the Mohicans BY JAMES FENIMORE COOPER
The Legend of Sleepy Hollow BY WASHINGTON IRVING
The Life and Opinions of Tristram Shandy BY LAURENCE STERNE
The Lost World BY ARTHUR CONAN DOYLE
The Man Who Was Thursday BY GILBERT K. CHESTERTON
The Master Mind of Mars BY EDGAR RICE BURROUGHS
The Mayor of Casterbridge BY THOMAS HARDY
The Memoirs of Sherlock Holmes BY ARTHUR CONAN DOYLE
The Merry Adventures of Robin Hood BY HOWARD PYLE
The Metamorphosis BY FRANZ KAFKA
The Mill on the Floss BY GEORGE ELIOT
The Murder on the Links BY AGATHA CHRISTIE
The Mysterious Affair at Styles BY AGATHA CHRISTIE
The Nature of Things BY TITUS LUCRETIUS CARUS
The Odyssey BY HOMER
The Old Curiosity Shop BY CHARLES DICKENS
The Origin of Species BY CHARLES DARWIN
The Picture of Dorian Gray BY OSCAR WILDE

The Pilgrim's Progress BY JOHN BUNYAN
The Portrait of a Lady BY HENRY JAMES
The Prince BY NICCOLÒ MACHIAVELLI
The Prince and the Pauper BY MARK TWAIN
The Problems of Philosophy BY BERTRAND RUSSELL
The Prophet BY KAHLIL GIBRAN
The Republic BY PLATO
The Return of the Native BY THOMAS HARDY
The Scarlet Letter BY NATHANIEL HAWTHORNE
The Science of Getting Rich BY WALLACE D. WATTLES
The Secret Garden BY FRANCES HODGSON BURNETT
The Sign of the Four BY ARTHUR CONAN DOYLE
The Souls of Black Folk BY W. E. B. DU BOIS
The Strange Case of Dr. Jekyll & Mr. Hyde BY ROBERT LOUIS STEVENSON
The Swiss Family Robinson BY JOHANN DAVID WYSS
The Tale of Peter Rabbit BY BEATRIX POTTER
The Tenant of Wildfell Hall BY ANNE BRONTË
The Three Musketeers BY ALEXANDRE DUMAS
The Time Machine BY H. G. WELLS
The Valley of Fear BY ARTHUR CONAN DOYLE
The War of the Worlds BY H. G. WELLS
The Warlord of Mars BY EDGAR RICE BURROUGHS
The Wealth of Nations BY ADAM SMITH
The Wind in the Willows BY KENNETH GRAHAME
The Wizard of Oz BY L. FRANK BAUM
The Woman in White BY WILKIE COLLINS
This Side of Paradise BY F. SCOTT FITZGERALD
Through the Looking-Glass BY LEWIS CARROLL
Thuvia, Maid of Mars BY EDGAR RICE BURROUGHS
Tom Jones BY HENRY FIELDING
Treasure Island BY ROBERT LOUIS STEVENSON
Twelve Years a Slave BY SOLOMON NORTHUP
Ulysses BY JAMES JOYCE
Uncle Tom's Cabin BY HARRIET BEECHER STOWE
Utilitarianism BY JOHN STUART MILL
Vanity Fair BY WILLIAM MAKEPEACE THACKERAY
Villette BY CHARLOTTE BRONTË
Walden BY HENRY DAVID THOREAU
War and Peace BY LEO TOLSTOY
What's Wrong with the World BY GILBERT K. CHESTERTON
White Fang BY JACK LONDON
Wuthering Heights BY EMILY BRONTË

Printed in the USA
CPSIA information can be obtained
at www.ICGtesting.com
LVHW092037091224
798732LV00021B/92/J

* 9 7 8 1 7 7 4 7 6 1 8 6 1 *